ON GIANTS' SHOULDERS

D0913691

Also by Melvyn Bragg

FOR WANT OF A NAIL
THE SECOND INHERITANCE
THE CUMBRIAN TRILOGY:
The Hired Man
Kingdom Come
A Place in England
THE NERVE
WITHOUT A CITY WALL
THE SILKEN NET
AUTUMN MANOEUVRES
LOVE AND GLORY
JOSH LAWTON
THE MAID OF BUTTERMERE
A CHRISTMAS CHILD
A TIME TO DANCE
A TIME TO DANCE: THE SCREENPLAY
CRYSTAL ROOMS
CREDO

Non-Fiction:

SPEAK FOR ENGLAND
LAND OF THE LAKES
LAURENCE OLIVIER
CUMBRIA IN VERSE (Edited)
RICH: The Life of Richard Burton

ON GIANTS' SHOULDERS

Great Scientists and Their Discoveries from Archimedes to DNA

Melvyn Bragg

WITH

Ruth Gardiner

John Wiley & Sons, Inc.

New York • Chichester • Weinheim • Brisbane • Singapore • Toronto

This book is printed on acid-free paper. 

Copyright © 1998 by Melvyn Bragg. All rights reserved

Published by John Wiley & Sons, Inc.

First published in the United Kingdom in 1998 by
Hodder & Stoughton, a division of Hodder Headline PLC

No part of this publication may be reproduced, stored in a retrieval system, or transmitted in any form or by any means, electronic, mechanical, photocopying, recording, scanning, or otherwise, except as permitted under Section 107 or 108 of the 1976 United States Copyright Act, without either the prior written permission of the Publisher, or authorization through payment of the appropriate per-copy fee to the Copyright Clearance Center, 222 Rosewood Drive, Danvers, MA 01923, (978) 750-8400, fax (978) 750-4744. Requests to the Publisher for permission should be addressed to the Permissions Department, John Wiley & Sons, Inc., 605 Third Avenue, New York, NY 10158-0012, (212) 850-6011, fax (212) 850-6008, email: PERMREQ@WILEY.COM.

This publication is designed to provide accurate and authoritative information in regard to the subject matter covered. It is sold with the understanding that the publisher is not engaged in rendering professional services. If professional advice or other expert advice is required, the services of a competent professional person should be sought.

Library of Congress Cataloging-in-Publication Data

Bragg, Melvyn

On giants' shoulders : great scientists and their discoveries : from Archimedes to DNA / Melvyn Bragg.

p. cm.

Includes bibliographical references and index.

ISBN 0-471-35732-4 (cloth : alk. paper)

ISBN 0-471-39684-2 (paper : alk. paper)

1. Scientists Biography. 2. Science–History. I. Title.

Q141.B777 1999

509.2′2–dc21

[B] 99-29005

Printed in the United States of America

10 9 8 7 6 5 4 3 2 1

To my son Tom
Good luck with the Science.

Bernard of Chartres used to say that we are like dwarfs on the shoulders of giants, so that we can see more than they, and things at a greater distance, not by virtue of any sharpness of sight on our part, or any physical distinction, but because we are carried high and raised up by their giant size.

JOHN OF SALISBURY, 1159

If I have seen further it is by standing on the shoulders of giants.

SIR ISAAC NEWTON, 1675

CONTENTS

INTRODUCTION

ON GIANTS' SHOULDERS focuses on twelve scientists who, in the last two thousand five hundred years, changed the world both as we perceive it and as we live in it. From Archimedes in Ancient Greece to Francis Crick and James Watson in mid-twentieth-century England, these landmark minds, their lives, their struggles, their colleagues and rivals are explored and unravelled by some of today's leading scientists. In combination, their stories and discoveries constitute a guide to the history of science.

There are many different ways to write history. Using the lives of those who were truly great has a long and honourable tradition. The advantages are many, particularly in the history of ideas and especially given that those ideas often crystallised in a single mind, whatever tributary contributions there might have been. A central figure can arouse an interest which leads us into ideas: an individual can typify as well as exemplify a sudden breakthrough in thought, a single figure can allow the context – the time, the culture, the place – to be brought to bear. The Greek historian Plutarch, who used this method, is still read today, and whatever the strength of the argument about the mass of interrelated movement which, like a broad tide, carries history forward, the attraction and sometimes the centrality of individuals remain.

Here the cast is formidable. Whether you look at Faraday apparently stumbling into science from the humblest and most unlikely beginnings, a bookbinder's apprentice become a man who commanded the intellectual society of London, or the heroic Marie Curie, arriving in Paris from Poland with little but the determination and brilliance which was to net her two Nobel Prizes and a place, literally, in the Pantheon; whether it is the impenetrable Newton of deep vengeance, even deeper religion and a force of invention all subsequent scientists look on with awe; or Galileo, the feisty Medician courtier, successful bringer of new worlds and unsuccessful intriguer in old ones; whether Poincaré, whose absent-mindedness co-existed with outstanding clarity over such a wide range, or Lavoisier, his countryman, guillotined for tax collecting yet reinstated as 'The Father of Chemistry', whether Darwin, Freud, Einstein, Crick, Watson or the man thought of by some to be the first mathematician and physicist, Archimedes – these are people with rare and often thrilling minds, as well as the makers of discoveries which changed the world.

Certainly there are very few of the two dozen and more contemporary contributors to this book who doubt the impact, the transfiguring force of intellect and often the sheer breathtaking originality of those discussed here. Several have reservations about the 'genius' theory of science and this is given full play throughout; but even those with the keenest reservations were unstinting in their appreciation of what a tremendous contribution each of these figures made.

The involvement of scientists who are alive now is fundamental to the book, and I found their contributions invaluable. When, in the Darwin chapter, for instance, you bring to bear the combined though often opposing forces of Stephen Jay Gould, Richard Dawkins, Daniel Dennett, Janet Browne, Richard Darwin Keynes and John Maynard Smith, a great range of points can be made with comparative brevity.

First, they get to the heart of Darwin's achievement rapidly, accurately and reliably. What he brought to the expanding universe of thought is described surely and succinctly. Second, they comment on the context out of which he grew – not only his personal background but also the intellectual and general cultural context of the day. Finally, they bring Darwin and his ideas into the 1990s and show how they have developed and been developed since Darwin's day: what the main thrust has been; what the chief disputes are; what potency has endured and, in Darwin's case, as in so many others, grown. It is a rich and an uncommon mix.

The technique of interviewing has a part to play here. By asking simple but, I hope, central questions, the essentials are described and the main points made. Because those interviewed are invariably so able, what is given in a fairly brief answer is often a brilliant encapsulation, and because the contributors carry such authority, these encapsulations are solid stones which rapidly build up the structure of the key ideas in question. Certainly in the shaping and linking and cross-cutting of these interviews, the aim has been to create in each case a portrait, a case study and a commentary all combined.

My own interest in science is, I would guess, fairly typical of my generation. I was bought a chemistry set one Christmas and mixed and bubbled as best I could, secretly hoping to bring about a fantastic transformation of matter or, at the very least, a creditable explosion. I was blessed with neither. Although I greatly enjoyed maths at school and at one stage wanted to take it in the sixth form, I was never enticed into physics. This is not to blame the teachers, for, soon after World War II and in a small Northern grammar school, teachers, especially it seemed in physics, came and went at some speed. Biology was fascinating but something about the class did not fire me. I ended up studying History, Latin and English in the sixth form and went to university to read History.

Science for me as a boy was the science of wonder. It was Dan Dare in the comic *The Eagle* with the massively brainy Mekon, his so much cleverer opponent. It was fascinating facts, and figures, also usually in comics or tucked away in popular magazines or newspapers. On the darker side, there was the Atom Bomb and the fears it provoked about Armageddon and, by extension, fears that scientists could destroy the world in so many ways every bit as easily as they could illuminate or improve it. Then, about ten years ago, I became aware that I was missing a great deal about the times in which I lived. Science was almost entirely off my radar. Yet there was a buzz about science which had not been there in the fifties, sixties and seventies. Perhaps it had been there, but I had simply not recognised it.

I began reading reviews and articles about genetics, about cosmology, about the mind, about matter, about, literally, universal questions and, equally literally, molecular questions, and I saw that I was missing out on a great deal. For a while I did no more than peep through the keyhole, sure that my lack of any serious science training denied me entrance to what increasingly seemed the most dazzling intellectual pleasure-garden of the late twentieth century. The world was being reinvented, reshaped, reunderstood in there. Was I to be for ever forbidden entrance just because I had dropped physics? It seemed a poor excuse to miss out on what, more and more clearly, seemed to me to be the defining and by far the most exciting thoughts in my time. To live through a period of intellectual ferment and not to be allowed to know what was going on seemed unjust! More to the point, it was silly.

Fortunately – or perhaps as a consequence of the increasing fervour in the sciences – books began to emerge which unashamedly sought to engage the general reader. We might not understand the maths, or the physics, or the chemistry, or the statistics, but this did not matter. The explanations were often clear enough for us to hang on to something of the substance. By 'us', I mean non-scientists, like myself – and there proved to be many of us. Of course none of us could have tackled an A Level paper in the sciences, even after a couple of years of steady reading of those books, but at least we – or at least I – no longer felt left out. The chief way in which the later twentieth century described itself was at least approachable. I am sure that quantum physics can be simplified out of all real meaning and that the chemistry of DNA still requires more than is in all my knowledge of biology, that chaos theory is much easier to get wrong than right after however many books and articles and that the Big Bang, facile now as a phrase, conceals a continent of numbers way beyond me. But, despite all of that, I was somehow present, I thought; I was not a player and never would and never could be. But I was at the game.

So what was the game? It seemed to me that over the last century science has gathered together the forces of the last two hundred and fifty years and put itself into the command position in our intellectual and perhaps even our imaginative economy. Ideas are flung out, sprayed out like sparks from a welder's torch. Ideas which not only take us on trips to the beginning of time but which take us on trips to essential matter and everything in between. New patterns, relations, synergies, analyses and symmetries, new worlds, new words, newness swarming in at every pore: this was the world humankind had arrived at and this was the world gathering force, ever more sure of itself, it seemed, ever more dominating, with science turning into technology like magic and technology turning the earth into a new planet. Better? Worse? Doomed? Released? Who really knew? But the speed and the whoosh of the enterprise is wonderful.

Learning about science, for me, had the effect of transforming the world. Sometimes the brain seemed out of control as it sped to the edge of the universe and then suggested, quite calmly, a parallel universe, or dwelt on knowledge which was hair-raising – such as that physicists were within the merest fraction of a second of identifying the first moment of time and space. Equally, where was this probing of mind taking us, this mapping of the genes and the penetration of the micro-world every bit as astonishing as those essays in space? Outside all of this, the absorbing, mundane, diurnal world of bringing up a family, forging friendships, falling in and out of love, seemed to go on unaffected by the sound of science; but how long before that too was touched by its influences?

An advantage I had was that I present a weekly one-hour discussion programme, *Start The Week*, on BBC Radio 4. Over the last eight years in conjunction with producers, principally Marina Salandy Brown, Ruth Gardiner and Olivia Seligman, I have been able to meet and interview many of those – British and American in particular – who write the books, who do the work, who know the field, who can and, thankfully, want to reach out to all those interested, even if those who are interested are ignorant. It would not be too strong to speak of many of today's scientists having something of an apostolic mission. They want to explain, at least, the wonders of their worlds – and if they make converts, so much the better.

Scientists began to appear on *Start The Week* in greater and greater numbers until they formed the most numerous 'block' of guests on the programme. Coincidentally, but not, I think, at all accidentally, the audiences grew, as did their response to the scientists. There were always many more phone calls asking for the titles of the science books discussed than any other, more letters, more comment in the press and more anecdotal evidence that some sort of need was being met. My assumption was that many of the *Start The Week* listeners were rather like myself – thirsty for information in huge territories of knowledge from which they thought they had been excluded. They were grateful to be let in, especially when the gates were opened by men and women at the forefront of present research.

After several years, the notion of a science series on Radio 4 came up, one in which I would play the part I played in *Start The Week*, that is to interview leading scientists from the perspective of a non-scientist with a serious interest in what was happening. We sought an overall idea which could be written and assembled to produce a story rather than just a cluster of interviews.

I went for the simplest notion – homing in on a dozen great figures and assembling the programmes around them. Obviously others could have been included and many are mentioned in passing – like Kepler, Maxwell and Copernicus – as well as scores of co-workers and equal labourers in the field. All could have been brought more firmly into the foreground. But we stuck to our original idea and, by enlisting the generous help of so many fine contemporary scientists, enriched it far more than I had ever thought possible.

This, then, is *On Giants' Shoulders* – a modest enterprise but one which seeks to reach out to the deep past of science and also pin it to the present day. The giants are as clear as pylons striding down the landscape of history. Supported by them and discoursing on them are many current scientists who themselves could be called to gianthood. Supported by *them* is this non-scientist.

There are several acknowledgments which must be made.

The first is to all the scientists whose generosity in time and patience gives the book whatever qualities it has.

The second is to the BBC, especially Radio 4, and particularly Ian Gardhouse, who took on the radio series that was the begetter of the book.

The third is to Ruth Gardiner, the producer. Very properly, she shares the title page but I would like here to pay tribute to her tremendous skill and persistence in organising and meshing together a series on which we both learned a great deal. That she did it while producing a more important work (since arrived and called James), makes me admire her even more. Helping her over the year were researchers Karen Holden, Jeanette Thomas and Alice Cooper. It was a dedicated and first-rate team.

I am also grateful to Ian and Margaret Millar for reading the typescript, to Claire Squires at Hodder & Stoughton, and to Carole Welch my editor, who has taken on this sweep of work new to both of us with her usual scrupulous zest.

I have added to my comments and reorganised the material for book form. The book differs from the radio series principally in the ampler amount of material contained in the book. It has been possible for me to return to the original transcripts of the interviews, mercilessly pruned for the thirty-minute radio programmes, and include much more.

It has been a marvellous enterprise.

Archimedes

(C. 287 BC–212 BC)

c. 287 BC Born Syracuse, Sicily.

Archimedes' extant works, published in the form of correspondence with mathematicians of his time, are *On the Sphere and the Cylinder*, *On the Measurement of the Circle*, *On Conoids and Spheroids*, *On Spirals*, *On the Quadrature of the Parabola*, *On the Equilibrium of Planes*, *On Floating Bodies*, *On the Method of Mechanical Theorems* and *The Sand-Reckoner*. Many other works by Archimedes referred to by later mathematicians such as *On Spheremaking* are now lost.

c. 212 BC Killed by a Roman soldier during the siege of Syracuse.

75 BC Cicero finds and restores his tomb.

The First Scientist?

> Archimedes is so clever that sometimes I think that if you want an example of somebody brought from outer space it would be Archimedes. Because he, in my view, is so original and so imaginative that I think he is better than Newton. Whereas Newton said 'I have only seen so far because I have been standing on the shoulders of other giants', there was nobody for Archimedes, nobody's shoulders for Archimedes to stand on. He is the first physicist and he is the first applied mathematician. And he did it all on his own from nowhere and I think that is just so amazing. It is breathtaking for me.
>
> Lewis Wolpert

THIS STORY of science begins with a legendary figure who appears to have conjured scientific thought out of thin air, more like a magician or a very great artist. Today's view of what a scientist does is partly that of a builder on the blocks of others – Newton's phrase speaks more persuasively to us. Yet Newton may well have been playing the clever game of false modesty: he was a man most careful of his reputation. He too, according to observers, admirers and followers, conjured discoveries out of the air.

Of course there were thinkers before Archimedes, and there had been great technologies for thousands of years, but science, that specific, abstract, even peculiar way of thinking, does seem to have arrived in Greece at the same time as so much else that has driven the intellectual cylinder of Western civilisation. We know now about the scope of mathematical activity in Mesopotamia between 1800 and 1600 BC and no one today fails to recognise the contributions of the Egyptians and the Babylonians and the Chinese. Yet for Lewis Wolpert, Professor of Biology as Applied to Medicine at University College, London, and a man of science whose views are often emphatic but always deeply informed, it was the Greeks who crystallised science.

> I am not a historian and I take it from other historians, but if you look through the books and you look at the work of people like Geoffrey Lloyd in Cambridge, and if you look through recorded history, the first recorded statements that one could regard as science come from the Greeks. Now there may have been some people around who did wonderful physics, but there is absolutely no record of them whatsoever.

I asked Lewis Wolpert how he was defining science.

> I am always very nervous about defining science, because I do not think that is the right way to go about it. You can give some of the characteristics of science – that is much better. It is really about looking at underlying principles. It is about understanding. Thales, in 600 BC – whether it is apocryphal or not does not matter – stood back from nature and tried to understand how it really worked. Without any relevance to application or technology, just a genuine curiosity – how does this all work? So when Thales said 'I think everything in the world is made of water in different forms', well, that is a scientific idea. He may be wrong, but that does not matter, it is a way of doing science.

Did he believe, I asked, that this was a real beginning, that other civilisations had great technologies – and we now know a lot about Chinese technology around the same time, for instance – but had no science? I wondered whether Chinese experts would dispute this with him.

> It is hard to tell. I take my view about Chinese science and the Chinese not having science of the kind we have had from Needham, the author of *Science and Civilisation in China*. If you read what Needham actually says, the great puzzle is why the Chinese did not do science, and part of the reason – which Einstein pointed out – is they did not have geometry. And the great triumph of the Greeks

> was that they did have geometry, and that was an enormous help to them. The Chinese science was rather mystical, and that is my reading of Needham. As I say, the great debate is why they did not get science – their technology was astonishing.

This led me to ask about something that intrigues me: if the discovery of science had not happened in Greece, would it not have happened at all? Did he think we could have missed that part of our development?

> I believe so. I do not believe that scientific thinking is something essential for human culture. And most cultures did not have it. Technology is quite different.

The distinction between technology and science is clearer in the mind of Lewis Wolpert than it is in that of others. Yet it carries conviction. The next question, then, is whether the Greeks invented geometry or whether they inherited it from the Egyptians and others.

> I would say they invented geometry and that when Thales said, or was supposed to have said 'All circles are bisected by their diameter', I am sure that if you had gone to an Egyptian and said 'Look, if I draw a diameter et cetera' the Egyptian would have said 'Yes'. But putting it in the statement 'All circles are

bisected by their diameter' – that is the beginning of geometry. Now these are general statements, not particular ones. So I would say the Greeks did invent geometry, yes. It was an approach to understanding the world for which there is no evidence that anybody else had.

There is a big discussion as to why the Greeks did science. Was it the open nature of their society? Was it the fact that they had leisure? There is a wonderful statement from Aristotle saying that people turn to this way of thinking or philosophy when they have time on their hands. So you needed to be moderately rich. I think it is all these things in some peculiar way. Also I think the Socratic method, this close questioning and answering, not letting any assumption go unchallenged, was probably absolutely crucial.

I would say that one of the characteristic features about science that distinguishes it from our day-to-day thinking is the requirement for internal consistency. Now we do very well with common-sense, but my guess is that a lot of our ideas if you examine them closely are contradictory. You cannot be like that in science. The fact is that the Socratic method really questions what your assumptions are. If we look very closely at everything, do any two of them contradict each other? Then we will have to say something is terribly wrong. And that is a very important feature.

It seemed then, from what Lewis Wolpert was saying, that what emerged in Greece was the invention or discovery or arrival of geometry and the pursuit of vigorous, close, logical discussion.

> And Aristotle, of course, with his ideas about logic. His idea of setting up a series of postulates and then drawing conclusions, that was monumental. You could also say that the formalisation of geometry (Euclid did not invent geometry, but as far as I can see he formalised it), the idea that you set up these postulates – a straight line is the shortest distance between two points and so forth – and then draw conclusions from them, that was absolutely Artistotelian, as I understand it.

Euclid (c. 300 BC) wrote *Elements of Geometry* in which he tried to bring together all the mathematics known before him and his own important work on geometry. He attempted to set up a system of proof which was to influence mathematics for two millennia. Aristotle (384–322 BC) described a philosophy of science and was preoccupied with the principles of logical thought. Sir Geoffrey Lloyd, who is Master of Darwin College, Cambridge, Professor of Ancient Philosophy and Science and a renowned expert on ancient cultures, has written that '*One of Aristotle's fundamental contributions was that he both advocated in theory and indeed demonstrated in practice, the value of undertaking detailed empirical investigations*'.

Archimedes was born in Syracuse in Sicily about 287 BC, the son of an astronomer. It is permissible to assume that he was aware of the importance of precise measurements from childhood, and aware also of the bank of knowledge – Egyptian, Babylonian as well as Greek – which could have been available to an educated household.

Archimedes was regarded, in his lifetime, with great awe. The Greek historian, Plutarch, writing about him some three hundred years later, commented:

> *The fact is that no amount of mental effort of his own would enable a man to hit upon the proof of one of Archimedes' theorems, and yet as soon as it is explained to him he feels that he might have discovered it himself. So it is not at all difficult to credit some of the stories which have been told about him. Of how, for example, he often seemed so bewitched by the song of some inner and familiar siren that he would forget to eat his food or take care of his person. Or how, when he was carried by force, as he often was, to the bath for his body to be washed and anointed, he would trace geometrical figures in the ashes and draw diagrams with his finger in the oil which had been rubbed over his skin.*

It is fascinating that the description of the devoted scientist as someone 'other-worldly' – so powerful an image in this century from the stories of Einstein to the dramatisation of *Dr Who* – begins here. Once again, comparisons with other undeniably great scientists are remarkable. Isaac Newton's concentration on his subject was, similarly, so profound and intense that he would forget the more mundane activities of life. In the case of Archimedes it seems to have been so extreme – according to myth – that he did not spare the time to wash or look after himself physically. Another parallel here is with the early saints of the Christian church, such as St Cuthbert, the extraordinary seventh-century saint whose powers both in his own time and for centuries afterwards were given the respect afforded scientific geniuses now. Cuthbert would not wash, his nails grew to excessive length, everything was neglected that would take away the smallest amount of energy from the obsession which possessed him.

Yet such picturesque details can obscure the plot. Archimedes appeared eccentric but, perhaps more accurately, it was the exhaustion and exhilaration caused by his preoccupations which made him oblivious to the norms of the workaday world.

Geoffrey Lloyd is also keen to ensure that the entertaining idiosyncrasies in the behaviour of the world's first physicist and first applied mathematician do not obscure his achievements.

> He is brilliant. He is clearly one of the most original, he is *the* most original Greek mathematician. So let us give him credit where credit is due. But that does not really explain much. In fact, it does not explain anything, because it just says 'Wow! What an incredible guy!' Well, he was an incredible guy, but my life's work is trying to understand how these things happened. It is not as if he was

> just the one in Syracuse and there was no one else elsewhere. Of course his particular contributions are remarkable, but that is always the case. It is the case in philosophy, it is the case in literature. But it is not as if we could explain Archimedes. Can you explain you, or can you explain me? Of course not. There is an element of total individuality and originality, thank goodness. But we do not want to be mesmerised by that, without diminishing the old codger's contribution.

Yet even Geoffrey Lloyd has to smile when he calls him 'the old codger'. Most people remember Archimedes for two things: for shouting 'Eureka!' in the bath when he discovered the law of buoyancy – what we now call the Archimedes' Principle – and for inventing the Archimedean screw, which is still used today in irrigation, conveyors and even in the old-fashioned mincer my mother used to have. From the theoretically astounding to the outstandingly practical.

He also made a model planetarium and designed powerful siege engines which kept the Romans out of Syracuse. To scientists he is important for his work on statics – the forces such as levers and weights – and hydrostatics – how forces act on bodies in water. In mathematics he worked out how to calculate the areas and volumes of spheres and cylinders, came up with a good approximation of pi and suggested a way of writing very large numbers.

There is little doubting the brilliance of his achievements. Lewis Wolpert has no doubts at all.

The queen of science, or the two queens of the sciences, are mathematics and physics, and what he did was to apply mathematics to physics. If you ask what is the triumph of science, it is mathematical physics, it is applying mathematics to the real world.

He took what appears to be a very simple problem, that is the balancing of weights on a beam. Now if you take a child and you say 'Will you please balance these weights on a beam; this one is heavy and this one is light', they will put the heavy one near the fulcrum and the light one a bit away. You know everyone knows that is the way. What Archimedes said is the ratio of the weights that you put on the beam is inversely proportional to their distance away from the centre when they are balancing. But what he did was to set out to prove it mathematically. And the proof is extremely ingenious; it is a little complicated – I think A Level students would get it, but it is not easy.

One of the key concepts in Archimedes, and that is remarkable, is the concept of a centre of gravity. In other words, you could represent the weight of several objects, or even of a large object, by a single point which represented the points from which the weight acted. Now that concept alone should have won him a Nobel Prize because this was a totally original thought, the idea that there was a centre of gravity for an object. And then the way he proves it, as I say, is a little complicated. He puts the heavy weight on the beam, then he puts on the light weight, and he says 'Let us pretend they balance'. Then what he does is, he divides them up into smaller units and by a clever geometrical trick he distributes the little units along the beam, so the centres of gravity of the units are the same. But when you look at the final result, there are as many on the left-hand side of the fulcrum as there are on the right and, therefore, he says, under this condition of course it will balance. Absolutely amazing.

Galileo called him divine, and there is a quotation from Galileo in which he says 'Look, without Archimedes I could have achieved nothing'. It is the approach of Archimedes; it is using the mathematics to try to understand the physical world. Now, of course, some people may say 'Oh, Galileo only said that because in the Renaissance people thought it was fashionable to praise the Greeks'. I do not believe so. Archimedes was so original and so amazing that when Galileo said that, I think he meant it.

Then, of course, his other approach was the floating bodies. Once again it is 'Wow!' He goes along to fluids and he makes the sort of assumptions that modern physicists make. He did not understand the concept of pressure, to be absolutely honest, but he had an idea that there were forces pushing things, and he describes this in such an elegant way. Then he considers when you lower something, say put a body into the water, he considers the pushing forces and explains why certain bodies will float and certain bodies will not, and why there is a loss of weight when you put something into a fluid.

I could give you a quotation from Archimedes, if I may.

Let it be granted that the fluid is of such a nature that of the parts of it, which are at the same level and adjacent to one another, that which is pressed the less is pushed away by that which is pressed the more.

And so forth. The details are not that important, but it is the language, it is the rigour, it is the simplification, it is laying it down. He says a little later in one of his postulates:

Let it be granted that any body which is thrust upwards in the fluid, is thrust upwards along the vertical drawn through its centre of gravity.

From out of space! It is a totally original way of thinking about what is happening when you immerse a body in fluids.

He also solved the problem of the crown, you know. That is the thing when he leapt from the bath shouting 'Eureka!' The King said 'Look, is this wreath made of silver or is it gold?' And Archimedes

> understood the concept of specific gravity, that is, for the same volume or the same weight, let us say, of silver and gold, the gold will occupy less volume than the silver. So all he had to do was measure whether the crown corresponded more with silver or more with gold by simply weighing it and then seeing what its volume was. Amazing. I keep saying – amazing.

Lewis Wolpert's compulsively enthusiastic use of the word 'amazing' take us back to the crucible of this revolutionary working of the mind. Geoffrey Lloyd is not as resolutely Archimedean as Lewis Wolpert.

> The important thing is to see that science is not a single phenomenon that suddenly springs fully armed from the head of Archimedes or anyone else. Of course, what Archimedes is doing is stupendously original in statics and in hydrostatics, for sure. One thing that I would take issue with Lewis on is that I think he has got too restrictive a view of science. If I can make an obvious point, if we have mathematicians who are able to develop eclipse cycles that enable them to predict lunar and solar eclipses, what on earth are we going to call that if we do not call it science? This seems to me a perfectly good example of science. Now, if that is the case then we have to include not just the Chinese, who are rather good at eclipse cycles, but

also the Babylonians – they started that particular investigation and did remarkably well at it. I think we have to spread the net quite wide and take it field by field.

I do not like talking about the beginning or the origin of science as such. For example, you can identify the first person to engage in the dissection of human bodies, you can identify people in Greece and China, what their programme was, what their agenda was. We can investigate and we can see that the programme was in fact something rather different.

The Chinese are brilliant technologists and they are far more interested in technology than the Greeks are. We cannot deny that they are brilliant technologists, but they are not *just* technologists. I mean – look at the mathematics. Liu Hui, third-century AD mathematician, who comments on one of the two main mathematical classics. Liu Hui engages in his own right in problems such as the approximation of pi, the relationship between the circumference and the area of the circle, and to determine the volume of a pyramid, for which you have got to use methods that are other than ordinary plane methods. Liu Hui goes into this, the way that you get a figure for pi, the circumference/diameter relation, that is exactly the same as the Greeks used.

What Archimedes is doing is stupendously original in statics and hydrostatics but, at the same time, there were models for him to draw on, because Archimedes has in the background a lot of mathematics. There is Euclid, and other mathematicians too, developing the most important notion you have in Greek mathematics, namely this notion of Axiomatic Deductive Demonstration. The point about demonstration is that it comes in different forms and you can prove things in different ways. You can prove things by just testing them and seeing they are OK.

The difference with Axiomatic Deductive Demonstration is that you derive the results of a whole field from a small number of basic primary instruments, the axioms, which you take to be self-evident and true, and from them you then proceed to develop, in the case of Euclid, the whole of Greek mathematics. The great feature of it is that it provides a certain kind of certainty, or incontrovertibility. That is where they really scored, that is what drove them, that was really what the prestige of mathematics consisted in. As a synthesis of early mathematical knowledge it is really quite remarkable, it took an awful lot of planning, an awful lot of synoptic imagination to get it in the form that he did. It is not just the mathematicians, the philosophers are concerned with the question of certainty too.

Another thing that I would quarrel a bit with Lewis over is that his view of Archimedes, although

> his enthusiasm is infectious, still savours a bit too much of the Greek miracle, you know – and suddenly there was a genius.

For a non-scientist to try to be umpire here is absurd and yet, although the subject is clearly debatable, I find that I side with Lewis Wolpert and see the observation of eclipse cycles, for instance, being less to do with science (although it may have provided a leaping-off point) than technology, although Geoffrey Lloyd would not agree. And I further side with Lewis Wolpert in believing that there is a vital distinction between the two, although, like Geoffrey Lloyd, I am enough of a twentieth-century man to be worried by the sudden arrival of genius. And yet, why not? There were many Elizabethan poets and playwrights but surely the best of Shakespeare is far beyond anything that had gone before. You could say the same of Mozart. Whatever the support system, suddenly there can be genius.

Geoffrey Lloyd agrees that there was something special about the way the Greeks did their science. What is important, according to Lloyd, is the way that Archimedes and others developed methods of proof. It is an oddly modern explanation: competition coupled with the intellect.

> The competitiveness is undeniably there, and that is something they all go in for, and they have to, because how did they earn their living? We have got detailed evidence about both philosophical schools and medical schools, where they slug it out with

> one another, both in terms of particular theories and in terms of what those theories are based on, the methods. How can we know the unseen causes of diseases in the body, for example? There would be an argument about that and a lot would hang on it because pupils would see, would participate in a public debate. It sounds rather negative, it sounds as if their methodology is just for point-scoring. But it is not just for point-scoring because they produced some fantastic ideas. It is not as if it is just brilliant reflection on what you need for science. There is much more to it than that and it takes you back into 'I am doing it this way, you are doing it that way; I am right and you are wrong'.
>
> This is miles away from anything you can find in China, because what you are doing in China is not engaging in public debate, least of all public debate that led to loss of face. What you do is you have an idea, you develop the idea and you say 'Emperor, my idea', and you hope that you can persuade the Emperor, and it is a pretty high-risk operation, because if you get it wrong, the fate of intellectuals in China is pretty gruesome. Lots of castration, lots of people being killed – it is very, very gruesome. Nevertheless, the Emperor was the one that counted, it was not persuading your peer group.

Geoffrey Lloyd especially admires the Greeks' rigorous method – springing originally, perhaps, from Socrates and the dialogues so influentially reported by Plato. This, he says, is their legacy across the centuries.

> They do not have scientific method in the way we have it. And they were not taught scientific method at school because there were no schools that were set up to train the next generation of scientists. The situation is very different from our current situation, but it shares one feature: namely, that if you actually talk to scientists they recognise there are many ways ahead or components of research that will help them to get the results they need. Listening to them talk, scientists in the laboratory or at lunch, you can see how miles away it is from the formal official presentation of the results which, when they publish them, they publish in a very impersonal way. Here we have a paper in such and such a journal and it is as if no one were there, it is as if everything were doing it by itself. But, of course, the scientists have set the thing up. Now actually that impersonality is to some extent already present in Greece, that is to say they want to depersonalise the results; at any rate in this tradition of the kind of science that aimed at certainty.

The most intriguing question of all is, simply, why did Greek scientists choose to do science? How and why did Archimedes turn his attention to floating bodies, or decide to find a new way to measure curved areas and volumes?

What did Archimedes and his contemporaries think they were doing? Once more into the breach, Professor Lewis Wolpert:

> To me it is a total mystery. Science did not help anybody, as far as technology was concerned, until really the late eighteenth century. So why science survived at all, I do not have a good explanation for. Maybe it was just intellectual satisfaction, maybe it was the praise. I mean, what do we scientists thrive on? Not money – yes, of course we would like money, but we thrive on the praise and admiration of our peers. That is our currency: praise and admiration. We do not always get it, but that is the currency. And maybe this was a society which did admire this and maybe that is why it persisted.

I quoted Lewis Wolpert back at him. He has written that '*Those who think all scientific ideas are merely temporary explanations to be replaced later should reflect on his theory, which will be right for ever*'. 'His' being, of course, Archimedes.

> That is true. I cannot bear the relativeness of people who go round saying 'The point about science is it is all transitory'. I suppose in the very strict sense, Archimedes was wrong, because if you did the quantification and used Einsteinian physics, yes, I suppose there would be another way that you would describe it. In the same sense that Newton was

> wrong. But he was basically right, you know. If you want to know whether a body is going to float or not, Archimedes' Principle will tell you for ever, it is like water is H_2O for ever.
>
> Of course, there are things changing in science and that is the essence, that things can change. If the evidence is there you change. It is painful, but you have got to change; but some of the key ideas remain for ever and I think Archimedes' ideas on floating and the levers, those I can guarantee are for ever. Maybe if you are travelling near the speed of light and therefore you are into an Einsteinian world, I am out of my depth, maybe it will not hold, but in the world which we inhabit, Archimedes is right.

This brought to my mind something else Lewis Wolpert has written: that there is a distinction between scientific ideas and progress and, let us say, artistic ideas and progress. Had Shakespeare not written *Hamlet* no one else would have done so, but had Archimedes not made his discoveries and inventions someone else would have. From what Wolpert said, however, it seemed to me that he was denying this. Characteristically, he tackled the dilemma head-on.

> I am in a cleft stick. You are absolutely right. My position about science, in general, is that given a community of people who are working on it, everything will be discovered. If not Einstein,

> William and Lawrence Bragg or whoever. And we know that simultaneous discovery in science is an everyday phenomenon. But maybe Archimedes and Aristotle and that lot really were different, because there was no science, they really started it all off. Once it got going, yes, the great giants speeded things up, Newton, Galileo and so forth. But once the idea got off the ground, everything will be discovered. The point about science is it is about progress. You cannot talk about progress in the arts, it has no meaning whatsoever.

How would Lewis Wolpert define a genius in science, I wondered? He had spoken of Archimedes as a genius. It is a word that we throw about and abuse but let us assume that it can have real meaning. How would it apply to Archimedes?

> He was different. He was a genius among geniuses, because for me he stands alone. Maybe I am being unfair, maybe the influence of Aristotle and the Greek astronomers is greater than I thought, because they were thinking about the cosmos and certainly Aristotle had ideas about the physical world. But the sort of rigour that Archimedes brought and the originality – certainly I would classify him as a genius.
>
> I think that in science if you take someone like Faraday or Darwin, they change the way we think

> about the world and you can either do this in dribs and drabs or someone could have this astonishing insight. It is claimed, for example, that Einstein's special theory of relativity would have come along anyhow, other people were getting there, but the general theory of relativity, that was almost Archimedean, it was from nowhere, it was just a totally original thought. Also what geniuses do is they found a field and they influence an enormous number of other people, they set up a tradition.

It seemed to me important, at the outset of this quest to convey something of the rich and complex history of science through a dozen of its greatest minds, that we took on such a fundamental debate. Today we are so used to, even brainwashed into, the notion that all scientific development comes from a group; that if one group does not arrive at the top of a scientific Everest then another surely will. It has, I think, gone from being a commonplace to a matter of faith, an ideology of the way theories and thought progress, even a theology as to how an understanding of the world is properly unravelled. The 'lone genius' versus 'the collective' is an argument which will, necessarily, run through a book such as this.

Dr Simon Schaffer, Reader in the History and Philosophy of Science at Cambridge University, is firmly opposed to Lewis Wolpert's notion of the lone genius. He believes that the history of science must not be interpreted as the story of an eccentric series of brilliant men and women.

> We can certainly describe people as geniuses, but I always ask myself, what are we doing when we do that? Historically speaking, for example, I think we simply do not know enough about the context of Archimedes. What we know about Archimedes biographically comes from a very small number of sources, written at a time when the notion of inspiration was the way one understood how supreme intellectual achievement happens.

I agreed that the lack of evidence made Archimedes a difficult subject and an easy target for such a view. But, I asked Schaffer, what about Galileo, Newton, Faraday, Einstein?

> I think all those are exceptional human beings. But I insist on the point that I really want to know what it is that we think we are explaining, what is it that we are capturing when we use these absolutely justifiable terms?

My response was that without these particular individuals such a line of scientific thought or investigation would not have gone ahead in the way it did. Schaffer had little trouble with that argument.

> That is exactly what I am sceptical about. It is very, very difficult indeed to run counter-factuals this way. No historian is particularly good at it. But

historians of science are especially bad at running counter-factuals, because it is very hard to unthink the knowledge of the world that we now have. But it seems to me that if we take a case like Newton, the shared techniques that Newton begins to develop, the extraordinary concentration of individuals who worked with him or competed with him or debated with him or picked up his work, is more striking than the unique perception he has – *if* what we are trying to explain is the pathway that the sciences have taken.

If our question, however, is psychological rather than historical, if we are trying to understand how it is that a particularly extraordinarily gifted individual does his or her work, then I think that is a question that we might discuss. But what I find less compelling is the privileging of the term 'genius' for just particular kinds of human activity rather than admiring – as I do, more than I can say, of course – these particular individuals for the achievements that they had.

The little we do know about the life and death of Archimedes suggests that he was killed by a Roman soldier during the siege of Syracuse in 212 BC, supposedly while he was working on a complicated equation. Plutarch describes the scene.

As fate would have it, the philosopher was by himself, engrossed in working out some calculation by means of

> *a diagram, and his eyes and his thoughts were so intent upon the problem that he was completely unaware that the Romans had broken through the defences or that the city had been captured. Suddenly a soldier came upon him, with an order that he should accompany him. Archimedes refused to move until he had worked out his problem, whereupon the soldier flew into a rage, drew his sword and killed him.*

Thus the story of Archimedes' life ends with the accent on obsession of such ferocity that it was, literally, fatal. It accounts, I believe, for the persistent linking, throughout cultures, of the deep thinker with the mad person, the utterly possessed artist, the completely engrossed child. Sometimes I think that whole literatures explaining obsession, genius, force of thought and invention have been inspired by Plutarch's sketch of Archimedes. But it *is* a very sketchy biography and much of Archimedes' work was lost to the West for centuries. It was only when Renaissance entrepreneurs realised they needed sophisticated mathematics to navigate the globe that the West again discovered his work.

Professor Lisa Jardine, Professor of Renaissance Study in the School of English and Drama at Queen Mary & Westfield College, London, spoke at the National Portrait Gallery in front of a painting of one of the great Renaissance instrument-makers.

> This is Nicholaus Kratzer, who is a mathematician and scientific instrument-maker. Holding his scientific instruments here, he represents a bridge

> between technical mathematics and applied science. You could not have navigation, for which he is making instruments here, without pure mathematics. The simple reason for the need for mathematics in the late fifteenth and early sixteenth century is that, once you turn a ship away from a coastline, you cannot find the coast again without advanced trigonometry. And advanced trigonometry was well developed by the ancient Greeks; the texts of Apollonius and Archimedes had been consistently and continuously studied in the Greek-speaking world and indeed in the Arabic-speaking world, where Greek was a perfectly familiar language, throughout the Middle Ages. We can see the results of that in the fact that Eastern ships were constructed using a far more advanced mathematical model for buoyancy than the West was capable of at the time.

Much of the credit for the rediscovery of Archimedes' work must go to Regiomontanus, the Renaissance scientist who was not only a brilliant mathematician but, luckily for posterity, also a gifted linguist. Following the fall of Constantinople in 1453, a few Greek speakers began to arrive in the West, bringing their ancient manuscripts with them. Regiomontanus not only understood the complicated mathematics contained within these beautiful books, he also set out to translate them, with the consequences Lisa Jardine explains.

> Regiomontanus was blown away by what he found in these manuscripts. We know exactly what he wanted to publish because he issued a schedule, a kind of advanced PR notice of the texts, and they include every important text – Archimedes, Apollonius and so on – and a whole series of proposed new commentaries and new translations. Sadly, Regiomontanus was struck down by the plague and died, but his assistant in the printing press continued to issue those texts. So Regiomontanus' scheme for mass-producing and mass-circulating the key Greek mathematical texts, in the original and in translation, was started in 1471 and continued until Copernicus used those very texts to produce his new world system in 1543.

Thus Archimedes leapt over sixteen hundred years to aid the discoveries which led with increasing rapidity to 'the modern world' – what could be called the world science has discovered for us.

From the Renaissance to the present day, scientists have looked back in awe at the mathematical and mechanical achievements of Archimedes. Many still remember him primarily as the man who jumped naked from the bath shouting 'Eureka!'. More than two thousand years after his death, that image of Archimedes vividly represents a moment of great scientific discovery. Does it matter if it is true or not?

To Professor Lewis Wolpert this entanglement of myth

and science – and this is not the last time it happens in the history of science – is deeply unwelcome.

> It irritates me intensely. Here is this genius. Now he may have got the idea in the bath. Maybe he was just like me, I would like to think I think a lot in the bath. But it somehow slightly diminishes the achievement. Perhaps I am being unfair, but it certainly irritates me. I like my heroes to be more dignified. I am probably being silly, but I do not think it is a story that helps terribly. He may have been thinking about it while he was in the bath, but it was not because he saw the water go up. That is nonsense. Do not believe a word of it!

Despite the irritation of Lewis Wolpert, there is little doubt that the legend will continue to be retold and continue to be attractive. For it says something profound – though in a way which can easily appear comic – about the way in which ideas seem to arrive, so suddenly, to some people. It can of course be argued, and convincingly, that such ideas have been long prepared for and that the work of others has been a major contribution. But – apart from the disputed point over Wolpert's claim that Archimedes was a sudden and sole genius – there *is* the Eureka factor. It may be irritating to some, but to others it is also human and true to their own modest experience.

Archimedes, with or without help from others, set the world on a course which has led it to the scarcely believable twentieth century feats of science. For centuries, however, his discoveries were almost wholly lost and played no part in a civilisation which continued to study mathematics, for example, but which took other aspects of thought, most notably theology, far more seriously than hydrostatics or even geometry.

Galileo Galilei

(1564–1642)

1564 Born in Pisa, Italy.

1581 Becomes a student of medicine at the University of Pisa, but does not complete his degree.

1589 Appointed a lecturer in mathematics at the University of Pisa.

1592 Becomes Professor of Mathematics at the University of Padua.

1600 Eldest daughter, Virginia, born to his long-term mistress, Marina Gamba. They have two more children, Livio and Vincenzio.

1606 Publishes *Operations of the Geometric and Military Compass*.

1609 Improves the telescope, invented in 1608 in the Netherlands.

1610 Publishes *The Sidereal Message* (or *The Starry Messenger*). Moves to Florence and the court of the Grand Duke of Tuscany as Chief Mathematician and Philosopher.

1612 Becomes member of the first truly scientific society, the *Academia dei Lincei*. Publishes *Discourse on Things That* Float on Water.

1613 Publishes the *Letters on Sunspots*, his first public pronouncement in favour of the Copernican system.

1615 Writes *Letter to the Grand Duchess Christina*, widely circulated but not published until 1636. Denounced to the Inquisition on charges of heresy and subversion.

1623 Publishes *The Assayer*.

1632 Publishes *Dialogue Concerning the Two Chief World Systems*. In October summoned to appear before the Inquisition.

1633 Tried by the Inquisition and forced to recant.

1638 *Discourse on Two New Sciences* smuggled out of Italy and published in Leyden.

1642 Dies.

The Columbus of the Stars

IN HIS day Galileo was called 'The Christopher Columbus of the Stars'. It was, it remains, very apt. But it is only a part of the story.

To the general public today, Galileo is probably better known for his persecution by the Roman Catholic Church because of his scientific theories than for the content and importance of the theories themselves.

Bertolt Brecht's play, *The Life of Galileo*, shows Galileo as first the champion of free, independent thought and then the craven capitulator, preferring mundane mortality to the eternal glory of standing by intellectual truth. The truth about Galileo is less simple and more intriguing. He brought to the heavens the Renaissance disciplines of science (newly informed by Archimedes via Regiomontanus), but his clash with the Church was more to do with the manners of Italian politics than the methods of re-energised science.

Galileo Galilei was born in 1564, in Pisa, Italy, the son of a musician. The association between music and numbers has an ancient history: in the sixth century BC, Pythagoras first discovered the connection between numbers and musical intervals (and like Galileo, he thought that numbers described the whole working of life) and it has been said of Mozart that, had he not been trained as a musician, he would have been a great mathematician. As with Archimedes, then, one can speculate that the first giant on whose shoulders Galileo stood might well have been his father; especially given that in the societies of Ancient Greece and Renaissance Italy, intellectual skills, like crafts, appear to have been held by and handed down within families.

After being taught by the Jesuits and, it has been suggested, considering becoming a priest – which, in his day, could indicate as much a shrewd worldliness as any religious calling – Galileo, even more shrewdly as it turned out, allowed himself to be enrolled in the School of Medicine at the University of Pisa. He left after four years before he had received a degree, owing to a lack of funds, but not before he had discovered a passionate interest in mechanics and mathematics – and he had a particular interest in Archimedes.

Galileo made an early living by teaching private pupils. But, besides being already a recognised scholar, he was always a clever crafter of opportunities for self-advancement and, by the age of twenty-five, he had become a lecturer in mathematics at the University of Pisa. A few years later he was appointed to the Chair of Mathematics at the more prestigious University of Padua, where he remained for eighteen years.

Mario Biagioli, Galileo's countryman, is currently Professor of the History of Science at Harvard University. He has engaged himself for some time with the difficult, often funny character of Galileo.

> He had a great sense of humour, definitely. And he was an arrogant man. He did not have much tolerance for disagreements but he was not somebody like Newton. Unlike Newton, Galileo was quite funny. He also had, I would say, a much more complex social life, especially the years he spent in Pisa and Venice. At that time he did not publish much at all, and so one assumes that he was more concerned with having fun than discovering the truth about the cosmos.

Father Michael Sharratt, Lecturer in Philosophy at Ushaw College, Durham, also has a well-informed view of Galileo's character.

> The thing about Galileo's character is that whether one likes him or not, he is fascinating. If one finds him really endearing, which he sometimes can be, if one can also enjoy his brilliance, because he was an absolutely brilliant teacher, populariser you might say, advertising copywriter, one cannot also ignore the fact that at times he was very savage when he was being attacked. Whether one would have liked Galileo if one had known him personally would have depended very much on sympathy with his ideas, I think, because he was – if one can misuse

> a word – a novelist: he really wanted to discover more and more novelties. His enemies by and large tended to be those with whom he had fallen out in controversial matters of philosophy or science, and since controversy or polemics was one of the routine exchanges of academic life – it still is – naturally one is going to find him in plenty of scrapes.

One of Galileo's attractions is that he performed experiments which any one of us could perform now. But it was he who first drew revolutionary conclusions that set off a shudder in the scientific cosmos, which has accelerated to seismic proportions. If you go to Pisa today – and it was there, in his home town, that Galileo made his first important discoveries – you find Galileo as a star of the tourist trail. Guides like Doctor Giovanni Giannini will escort you around the cathedral and explain something of the wonderfully childlike experiments that Galileo executed and, much more importantly, experiments of which he alone in his day perceived the consequences.

> This is the cathedral of Pisa. This is the cathedral where Galileo probably discovered the law of the pendulum. Probably he was sitting somewhere in the cathedral looking, watching an oil lamp swing. Apparently he measured each duration of these oscillations with his pulse beat. He must have had a very regular pulse beat indeed, he was still very young! Apparently, from the top of the Leaning

> Tower, Galileo performed experiments on the free fall of bodies – there is nothing written about it, we do not know whether it is a popular tradition. But we like to think of him from the top floor dropping these two balls of different weight from the same height, landing at the same time, to prove that two objects, the same form but different weight, fall at exactly the same speed.

'Maybe he did and maybe he did not,' commented Paul Davies. Davies, Professor at the University of Adelaide in the department of Physics and Mathematical Physics and a popular science writer, is sceptical of Galileo's need for experiments, though his admiration is, in its own way, just as enthusiastic as that of our Pisan guide.

> I think he provides one of the best examples of how the human mind, incredibly, is able to figure out how the world works, sometimes without actually going out and making the observation. Galileo had a certain style which we would now recognise to be that of the open-minded, free-wheeling experimental physicist. Most people go around with the intuition that heavy bodies fall faster than light ones and indeed they do in some circumstances. A feather will certainly fall more slowly than a lump of coal for example. But it was Galileo's genius to see that this was an incidental effect of air resistance. If you removed *that*, everything would accelerate equally.

> Galileo was able to deduce that bodies should fall equally fast without dropping things off the Leaning Tower of Pisa and he used the following curious argument: he imagined a light and a heavy body tethered together and asked the question 'Does the presence of the light body help or hinder the fall of the heavy body?' Now if, as Aristotle had maintained and, as intuition may tell us, heavy bodies fall faster than light ones, the presence of the light body would impede the fall of the heavy body, because it would tug on the rope, it would lag behind and restrain it. But Galileo reasoned that if we consider the entire assemblage of heavy plus light body together, well, that is heavier than the heavy body on its own, so the assemblage should fall faster, and so we arrive at the contradiction that the presence of the light body should both speed up and slow down the fall of the heavy body. That is obviously nonsense, and so the only way we can reconcile it is to suppose that both the heavy and the light body fall at the same speed. What a wonderful example of the power of reasoning to make sense of the world.

Paul Davies believes Galileo earns his place in our pantheon of giants because he was the first to use the techniques we associate with modern science.

> Galileo was able to apply human reasoning through the use of mathematics to the natural world and that

> was entirely novel. Today we take it for granted that all of science is founded upon the notion that there is a lawlike mathematical order in nature and that this order can be discerned through sufficiently clever and sometimes arcane procedures such as devising experiments where you isolate certain things, making certain careful observations and then doing a whole lot of squiggles on the blackboard. That is the way we do science now. But the fact that the world is structured in this way, the fact that reality is accessible to the human mind through the power of reason, and in particular the power of mathematical reason, is the absolutely crucial point. That is the scientist's world view and that was the world view that Galileo articulated so clearly.

Galileo expressed this idea in a key paragraph in a controversial work he called *The Assayer*.

> *Philosophy is written in this grand book. I mean the universe which stands continually open to our gaze. But it cannot be understood unless one first learns to comprehend the language and interpret the characters in which it is written. It is written in the language of mathematics and its characters are triangles, circles and other geometrical figures, without which it is humanly impossible to understand a single word of it; without these one is wandering about in a dark labyrinth.*

Although Galileo failed to establish the laws of motion (that was left for Newton), he did realise the importance of time: a fundamental breakthrough that has influenced physics to this day, as Paul Davies observes.

> This is the crucial thing. As we now know, the nature of motion is to be gauged by the passage of time, and we have concepts like velocity and acceleration that are now very precisely defined. These were just very vague concepts before Galileo – he made them precise. He showed that these were the quantities to concentrate on and I guess the most important part of his work in that area was to show that uniform motion is relative, purely relative. I can give you a very simple example. We have all had the experience of sitting in a railway carriage and suddenly you think 'We're off', and then you realise it is a train on a parallel track moving in the opposite direction. Uniform motion like that is relative. When I fly to England from Australia, most of the time I am in the air if I close my eyes I could not tell whether I am flying or sitting on the runway because it is uniform motion. Again this is a bit counter-intuitive and it certainly would have been in Galileo's time, when there were not any trains or aircraft. He spotted that uniform motion, though not accelerated motion, is purely relative. The very word 'relative' went on to become a fundamental part of physics and the theory of relativity which Einstein

developed in the early part of this century draws upon this Galilean principle. We still talk about Galilean transformations. So this experimentation and elucidation of the nature of motion really set the scene for Newton and generations of physicists afterwards.

Once again – as with Lewis Wolpert – we are back with a view of the uniqueness of science and its origins. At this stage I asked Paul Davies to sum up where Galileo stood in relation to science. Put at its simplest, could he justify being called, as he has been, the Father of Modern Science?

I think Galileo's most significant contribution was to draw together three disciplines previously separated – that is, mathematics, physics and astronomy. There was a feeling before then that astronomy was to do with the heavens, the very word says it all, belonged to a realm outside normal natural laws. Galileo famously turned the telescope on the heavens, and was able to deduce that the heavens were other worlds.

He spent a lot of time thinking about the nature of motion in the heavens and on earth and thought a common sense of principles obtained in both, and about that, of course, he was perfectly correct. Significantly, he also deduced that these principles would be mathematical in form. Ever since Ancient Greece the idea that the world was mathematical,

> geometrical, in number is something that was prevalent, but Galileo went beyond that. He saw that mathematics could be used to describe motion, in a way that did not totally tie down everything that had to happen. That is to say, it is impossible just from mathematics to know how the world is, we also need observation, the combination of experimental and theoretical science. This was the springboard from which modern science emerged.

It should be no surprise that this springboard coincided with a new individualism in the arts and with a view of life much more focused on the individual than had been the case under the clamp of medieval Catholic societies. The Renaissance marked the moment when European thought seems to have take a new breath and decided to become the intellectual and artistic power that was to sweep throughout the Europeanised world for the next five or six hundred years. Paul Davies sets science in this and its larger context.

> In my opinion what we call science draws upon two traditions which were very influential in what we call European Renaissance times. The first is Greek philosophy, human beings being able to understand the world by applying logic and reason. The second is the monotheistic religions, Judaism, Islam and Christianity: the idea that we live in a world which has been ordered in a rational way, by a lawgiver, that there

is a lawlike order in nature which is imposed from above.

Most cultures do not share that view. In many cultures nature is seen as a battleground, capricious, a tension – things pull this way and that – and what happens may be a compromise. But the monotheistic religions are very different. They assume there is a grand lawgiver presiding over all, a cosmic architect imposing an order on nature by fiat. This is not something which is going to drift around or be affected by the affairs of the world, it is a once and only deterministic abstract arrangement of things. Of course Galileo, like many of the early scientists, was deeply religious and, although he clashed with the Church, he believed God had imposed a lawlike order on nature and, in doing his science, applying his reason, he was glimpsing in some limited sense God's plan for the world. The motivation for doing science was this belief that there really is a scheme of things that can be discerned through experimentation. If he had not believed that, he would never have embarked upon science.

My belief is that the reason that science flourished in Western Europe is essentially that the European world view, from Greek philosophy and monotheistic religions, held that there was law in nature. It may not be apparent, but if you work hard enough it can be seen as a hidden subtext in nature, something to be dug out and displayed.

> It is interesting to wonder if those early scientists like Galileo ever believed that they would have a complete understanding of nature. Or whether that they could only apply these principles in a certain way. It was usual to say that man was created in God's image. I think they felt that the human mind reflected in some diminished way God's power, so there was some intellectual basis to nature. And of course scientists, even though most of them today are atheists, still have this world view. There is an intellectual view of nature that, by using our own intellects, we can read it out of nature and thus make new discoveries.
>
> I trace the modern scientific world view to Galileo not because he was the Father of Science but because he was the focus of these different strands of influence. He is undoubtedly a key figure.

What if Galileo had not lived, I asked? Davies's answer led him on to question the inevitability of science itself.

> If Galileo had not lived, I am sure that within a hundred years or so the ideas he came up with would have been found out by somebody else and the course of history may have been slightly different, but after a few hundred years I do not think it would have made a lot of difference. I think the absolutely crucial issue, the central point, is why did what we

now call science flourish in Europe at that particular
time, why not elsewhere in the world?

If say an asteroid had hit Paris in 1200 and
destroyed European civilisation, would science
ever have been discovered on this planet? I often
like to speculate that if there is intelligent life on
other planets that it may go on for millions or even
billions of years without them ever stumbling across
this extraordinary thing called science. We take it
for granted, we take the way we do science for
granted, but it is actually a very bizarre and arcane
set of procedures. This is not the sort of thing you
would naturally tend to do, and the fact that we look
back on episodes in history of the life of Galileo and
the other scientists very often leads us to take for
granted that it must have been that way. I think it
is anything but obvious, and I think the emergence
of science required a very particular world view
and a world view which is not shared by other
cultures, and may not be a world view which exists
if there are any intelligent creatures elsewhere in the
universe.

However important Galileo's contribution to understanding the world in a scientific way, it is not for this that most of us remember him, nor is it what brought him fame in his day. He would have remained a respected but rather obscure Professor of Mathematics were it not for the invention of the telescope by Hans Lippershey in Holland. In 1609, at the relatively late age of forty-six, Galileo learned about the new device. Within months he had improved the original instrument and made it immensely more powerful. Then he turned it on the stars. This is how the sight affected him.

Let me speak first of the surface of the moon which is turned towards us. For the sake of being understood more easily, I distinguish two parts in it which I call respectively the brighter and the darker. The brighter part seems to surround and pervade the whole hemisphere, but the darker part like a sort of cloud discolours the moon's surface and makes it appear covered with spots. Now these spots, as they are somewhat dark and of considerable size, are plain to every one, and every age has seen them, wherefore I shall call them great or ancient spots, to distinguish them from other spots, smaller in size, but so thickly scattered that they sprinkle the whole surface of the Moon, but especially the brighter portion of it. These spots have never been observed by anyone before me and, from my observations of them, often repeated, I have been led to the opinion which I have expressed, namely that I feel sure that the surface of the moon is not perfectly smooth, free from inequalities and exactly spherical as a large school of philosophers considers with regard to the moon and the other heavenly bodies. But that on the contrary it is full of inequalities, uneven, full of hollows and protuberances, just like the surface of the earth itself, which is varied everywhere by lofty mountains and deep valleys.

Galileo published his observations in *Sidereus Nuncius* (commonly known as '*The Starry Messenger*'), a little booklet which he dedicated, cannily but typically of the times, to that huge engine of patronage, the Grand Duke Cosimo de' Medici. The English ambassador of the day, hearing of Galileo's discovery, called it 'the strangest piece of news'. Father Michael Sharratt explains why.

> '. . . *the strangest piece of news*' because you could see that if Galileo was right, the whole trappings of the cosmological framework, which all educated people shared, seemed to be coming under fire. Whereas one had thought that all heavenly bodies including the moon were made of some perfectly heavenly materials, were perfectly spherical, even though people could see the man in the moon, if you like, Galileo said 'No, these have got to be mountains and valleys'. When he discovers four new planets, the four most visible satellites of Jupiter, again this was something unheard of, something sensational. So we find for instance that it makes an impact in England in the poems of John Donne, and perhaps the reason is that there was nothing so very difficult to understand. It was not very abstruse. The hard thing to accept was that the world was not as it had been thought to be. Difficult at first, of course, to trust this new instrument, this optic tube, this new telescope, but once one did, one could see with one's own eyes that things were more or less

as Galileo said and realised then that this was not compatible with the kind of world one had been brought up in.

I asked Paul Davies how Galileo's work on the movement of planets changed his contemporaries' view of the world. He had touched on this before, but it seemed to me important enough to develop.

Galileo made many famous discoveries by turning the telescope on the heavens. He became quite well known as an astronomer, but it is important to emphasise that before Galileo did his work the subjects of mathematics, physics and astronomy were separate. In fact astronomy did not exist. There were astrologers who worked for kings and princes, and so on, but scientific astronomy – the notion that the heavenly bodies obeyed the same laws as here on earth – was unknown, something that the Church frowned upon. One of the problems for Galileo was that the Church wanted to leave the heavens out of bounds as far as scientific enquiry was concerned. But Galileo, by turning a telescope on the sky, brought astronomy within the scope of science. He made many famous discoveries: mountains on the moon and spots on the sun. This was a great shock to religious people, because there was this perception by religious people that what was in the sky was the realm of God. Sun and moon were supposed

> to be perfect orbs, and if they had blemishes like mountains, then it made them look just like another world, rather than outside the domain of nature.
>
> He also discovered moons around Jupiter. Of course, Copernicus had already suggested that the earth was not the centre of the universe, that the sun was at the centre. But by discovering a mini-solar system around Jupiter, this was very strong evidence in favour of the heliocentric model. I think probably he was the first person to show that the domain of the heavens was something accessible to scientific enquiry, and the objects in the sky are other worlds as opposed to quintessential entities placed there by God.

With this discovery, and the careful cultivation of the Medicis, Galileo had ensured his route to success. Perhaps it seems strange to us to imagine a scientist at the heart of the Medici court, alongside artists and musicians. But Professor Mario Biagioli explains why Galileo had to become a courtier if his work were to be taken seriously.

> The very category 'scientist' or 'science' is something that we historians of science try not to use. Science did not quite exist as a profession in the Renaissance. You can talk about astronomers, you can talk about astrologers (most of the time these two are the same), you can talk about physicians, you can talk about surgeons, but the role of the scientist did not

quite exist. So the argument I have tried to make is that the princely court that often is presented as the headquarters of extravagance, as a place that has nothing to do with knowledge but has to do with spectacle, with etiquette, with intrigue and politics was, actually, an interesting institution for someone like Galileo, who wanted to establish a new professional role.

Galileo would be asked to debate with other philosophers in front of the prince or in front of their visitors. For instance, the debate on buoyancy started out apparently as a result of people using ice in their wine in the summer. People started noticing that ice floated. The question was – how come ice floats? So here you have a courtly fashion, ice in wine, that triggers a debate about buoyancy. Galileo and his opponents have to articulate the position and defend it publicly, and they have to do so quickly. The same way they could be asked when some comet becomes visible. Galileo could be asked – what are comets about? Where are they? A shooting star falls on Tuscany – they could say 'What are shooting stars?' So you have all these questions that are not necessarily connected to the professional interest of people like Galileo. Nevertheless they have to answer.

Spurred on by his success at court, Galileo's scientific discoveries were soon to draw him into religious controversy. In literature – most notably in Brecht's play *The*

Life of Galileo – and in the history books, Galileo has usually been seen as the martyr of science because he dared to argue that the earth went round the sun, not the other way round. As Paul Davies mentioned, he was not the first to do this. Copernicus had published his theory on his deathbed in 1543 in *De Revolutionibus Orbium Coelestium*, and there were other important astronomers like Johannes Kepler and Tycho Brahe working at the same time as Galileo who challenged the belief that all planetary motion revolved around the earth.

However, in 1632, against the better advice of some of his fellow scientists, Galileo wrote *Dialogue Concerning the Two Chief World Systems*. He wrote it in Italian so it could be understood by any literate layman and presented it as a discussion about the universe between friends, putting the old-fashioned arguments of the Greeks into the mouth of a character called Simplicio, a dullard, who was widely understood to be the current Pope. This time Galileo had gone too far. Mario Biagioli takes us back to the politics of the time.

> If you are a mathematician you can say ‘I think that Copernicus has given us a very interesting mathematical model to make sense of planetary motion’. Now that is a statement that would not put you in trouble with anybody because basically you would be read as taking Copernicus as a hypothesis. The problem was that people like Galileo did not stop there. They did not seem to say ‘This is a

hypothesis'. Galileo would say 'Well, there is very strong evidence suggesting that this is not a hypothesis, but the real thing'. That is when the claim becomes problematic. When you live in a country where the Church can write to your prince and say 'Please go and arrest Galileo and send him to Rome' – then you are in trouble.

Galileo had already been warned about upholding Copernicanism as a true doctrine not as a hypothesis. He was allowed to present it as a hypothesis but not as the real thing. That was in 1616. What happens in 1632 is that a friend of his becomes Pope, a friend who had written poems dedicated to Galileo praising his genius and so on. He asked Galileo to be careful, to write a book that was not a treatise, that was not like a dogmatic statement of the truth of the Copernican astronomy but rather a nice, playful dialogue. That is how it was supposed to be read. Instead it was read as a bad joke. It could have gone either way. I can easily think of the context in which the book would have been read as a philosophical comedy and people would have appreciated Galileo's literary skills and that could have been the end of the story. Instead, for intricate political reasons, it was read as a travesty.

The dialogue was published in February. By October, Galileo had been summoned to Rome and, after successfully delaying his appearance on the grounds of ill health, was on trial by

April of the following year before the Inquisition. Soon after, he publicly recanted.

I Galileo, son of the late Vincenzo Galilei, Florentine, aged seventy years, arraigned personally before this tribunal and kneeling before you, most eminent and reverend Lord Cardinals' Inquisitor General against heretical gravity throughout the entire Christian Commonwealth, having before my eyes and touching with my hands the Holy Gospels, swear that I have always believed, do believe, and by God's help will in the future believe, all that is held, preached and taught by the Holy Catholic and Apostolic Church. But whereas – after an injunction had been judicially intimated to me by this Holy Office to the effect that I must altogether abandon the false opinion that the sun is the centre of the world and immovable and that the earth is not the centre of the world and moves, and that I must not hold, defend or teach in any way whatsoever verbally or in writing the said false doctrine, and after it had been notified to me that the said doctrine was contrary to Holy Scripture – I wrote and printed a book in which I discuss this new doctrine already condemned and adduce arguments of great cogency in its favour without presenting any solution of these, I have been pronounced by the Holy Office to be vehemently suspected of heresy, that is to say, of having held and believed that the Sun is the centre of the world and immovable and that the Earth is not the centre and moves:

Therefore, desiring to remove from the minds of your

> *Eminences, and of all faithful Christians, this vehement suspicion justly conceived against me, with sincere heart and unfeigned faith, I abjure, curse and detest the aforesaid errors and heresies.*

The trial and recantation set off a quarrel between science and religion which has burst out repeatedly with the greatest intensity over the centuries. Although in Galileo's case the issue seems to have been a matter of presentation, it is difficult not to see the underlying conflicts ready to re-emerge. Still today some scientists are furious that religion makes claims to territories of knowledge in which it can show no 'proof', which religious thinkers counter by accusing scientists of being unable fully to explain away such fundamental notions as the origin of life, the pattern in life, even the Argument from Design, which holds that the very regularity and form of the universe 'proves' a single designer, a god. Yet there are those like the writer John Carey who deny any fundamental underlying conflict and see religion and science as two separate and valid ways of seeing truths about the meaning and source of life.

In 1633, the Church, once drawn into the argument, acted without equivocation. The Inquisition insisted that the *Dialogues* and all Galileo's existing works were to be banned. Copies of his books were burned, and Galileo was sentenced to imprisonment 'during the Holy Office's pleasure'. Within a year this was commuted to his permanent house arrest at home near Florence, where he was looked after by his daughter and was visited by such famous men as John Milton.

Despite the restrictions, Galileo started to write a new book, *Discourses on Two New Sciences*. Again written as a conversation, it contained the laws of mechanics which he had begun work on so much earlier. He smuggled a copy of it out of the country in 1638, four years before his death in the year Sir Isaac Newton was born. The house arrest was never lifted.

The science writer Margaret Wertheim believes we must not exaggerate Galileo's position as a martyr of science.

> We tend to have this picture in our minds that Galileo was imminently about to be put on the rack or burnt at the stake, but this in fact is not true. He was not some local village witch. He was a very famous man. His book *The Starry Messenger* had made him an absolute celebrity. In many ways he was the Stephen Hawking of his day. So this was a respected and powerful man, and throughout the trial he was kept in a Cardinal's palace, he was treated with dignity. Yes, the threat of torture was formally read out, but it was really more a formal proceeding. He was never really in danger. There was not anyone down in the basement oiling the rack or stoking the fire.

The argument is complicated further by the fact that Galileo never actually proved that the earth went round the sun, nor could he disprove Tycho Brahe's alternative theory that the sun orbited the earth while the other planets orbited the sun. Margaret Wertheim takes this point up.

> We have this idea in our society now that Galileo was some sort of caped crusader of the intellect, defending truth and justice in a scientific way. But the problem with the traditional mythology about Galileo is that he did not have proof at all. Nobody had proof in the early seventeenth century that the earth really was going round the sun. It looked like a reasonable hypothesis, but equally at the time you could believe in several other systems and there really was not definitive proof either way. In fact the Jesuits said they would actually revise their position if he could provide definitive proof and, given fifteen years, he simply could not do it, because nobody could at the time. Now the fact that he turned out to be right in the long run does not mean that the Church was obliged to accept what was simply a workable hypothesis at the time. So the Church has been given quite a bad rap in this case.

Father Michael Sharratt agrees that the role of the Church in the trial of Galileo was not a simple one. Galileo always thought of himself as a good Catholic. And Cardinal Bellarmine, the main representative of the Church and Galileo's key opponent, is no simple villain. Sharratt develops this argument.

> The heart of the difficulty was this question – could Galileo *prove* that Copernicanism was the true system of the universe? Underlining this was the shared idea that science is about proving things with absolute certainty, the Aristotelian idea of demonstration from necessary causes. Galileo himself tended always to be over-confident, to exaggerate his achievements and not make room for other people. In his more sober moments, he did know that, though he had not proved Copernicanism was true, he had certainly shown the serious defects of rival opinions. It was quite sensible for him to think that, if we can put it in terms we can accept nowadays, the Copernican idea was going to turn out to be the one nearest to the truth. But when it came to it, he could not deliver what Cardinal Bellarmine wanted: absolute proof. One snag about that, of course, is that we have to ask ourselves what would Cardinal Bellarmine have accepted as conclusive truth? These are the counter-factuals, or the ifs of history, that we can discuss with fascination, indefinitely.

Yet the trial of Galileo remains important in the debate – perhaps the battle – between science and religion. The Roman Catholic Church certainly felt so when John Paul II decided to re-examine the Galileo affair in 1992, more than three centuries after the trial. Father Michael Sharratt has studied this episode.

> The purpose behind this was not simply to right a wrong. It was the underlying thing that motivated Pope John Paul. He wanted to begin a process in which faith and reason, theology and science, could positively co-operate with each other. I think that was the aim of it and, given that, quite a good symbol of it was to reflect on the Galileo case, to praise him as the founder of modern physics and to say quite roundly that he made a better job of the theology than the professional theologians of his time.

The last word should go to Paul Davies, a physicist unafraid to use the word God in the title of his popular science books. He believes three centuries after Galileo that we must continue to draw careful boundaries between science and religion, and learn at least that lesson from the life of Galileo.

> The role of religion is to *interpret* how the world is, not to *tell* us how the world is, and so long as the Church was insistent on trying to dictate to scientists how nature is, it was bound to come into conflict with science. It did again, of course, with Darwin. Over the history of the Church as an institution, if you look back it seems to be a long-running sequence of retreats in the face of scientific advance. This has given a lot of people the impression that science and religion are perpetually in conflict and religion is usually the loser.
>
> But what I always say is you do not go to a

physicist and ask about moral questions or about good and evil. When you stop and think about what the role of religion is in society there are two really rather distinct roles. One is something about God, the grand architect of the universe, and then there is the other side of religion, which is how we should lead our lives, the issues of good and evil and personal responsibility and so on. My feeling is that the grand system of the world is something which is best left to scientists to work out. We can then go on and draw certain conclusions or interpretations about that as to whether there is a meaning or a purpose behind it all. But when it comes to human behaviour, human society, then I think we are a long way from science attempting to replace the role of religion in everyday affairs. I think it would be quite wrong for scientists to claim that they have some unique insight into questions of human morality, or good and evil. They have a contribution to make, most certainly, but you certainly do not go to a physicist to ask about a moral dilemma.

Sir Isaac Newton

(1642–1727)

1642 Born in Lincolnshire, England.

1661 Enters Trinity College, Cambridge.

1664 Writes *Certain Philosophical Questions*.

1665 Graduates and returns home to Woolsthorpe after plague closes the university.

1667 Elected to a fellowship in Trinity College.

1669 Becomes Lucasian Professor of Mathematics at Trinity College. Writes *On Analysis by Infinite Series*.

1672 Becomes Fellow of the Royal Society.

1684 Writes *On Motion*.

1687 *Principia Mathematica* published in three volumes.

1696 Made Warden of the Mint at the Bank of England, and made Master two years later.

1700 Named a foreign associate of the French *Académie des Sciences*.

1703 Elected President of the Royal Society.

1704 Publishes *Opticks*, waiting until after the death of his rival, Robert Hooke.

1705 Knighted by Queen Anne.

1707 Publishes *Arithmetica Universalis*.

1727 Dies, and is buried in Westminster Abbey.

Standing on the Shoulders of Giants

Sir Isaac Newton was born in 1642 at Woolsthorpe Manor in Lincolnshire. His early life on a working farm seems some distance from the pomp and ceremony that accompanied his state funeral in Westminster Abbey eighty-five years later. That event prompted Voltaire to say that England was honouring a mathematician as other nations honoured a king. Indeed, many of his contemporaries regarded him as a God and few today would disagree that he was one of the greatest scientists who ever lived. Even if the story of an apple falling from a gnarled tree (still, part of it, in the garden) leading to the discovery of the law of gravity is a myth, there is no doubt that it was Newton who was the first to explain the universe in the terms we understand and use today. Yet genius or not, the small boy who played with his intricate home-made models in the farmyard orchard became an arrogant and difficult man who fell out with many of the other great scientists of the day.

People are always interested in how someone of Newton's immense stature began his studies. It seems the farmhouse in the middle of the countryside was not as humble a background as later mythology would have it, but it was not a scientific background either. I visited Woolsthorpe Manor in the company of Dr Robert Iliffe, Lecturer in the History of Science at Imperial College, London, and asked him what Newton's childhood was like.

> Even before he was born (his father died three months before his birth) Newton was marked out as somebody very special. He said later that he only barely survived his birth. He was born on Christmas Day 1642 and, when he was three, his mother went off to live with the wealthy rector of the nearby parish and Newton was left on his own for the next eight years. We know very little about that, unfortunately, because for many people, of course, that is the key period of their life. He was brought up by his mother's mother, Granny Ascough and again we know very, very little about what happened then. We do not know how much contact he had with his mother, we know very little about his attitude towards his stepfather other than that he did not like him. We do know that at one stage later he said he wanted to burn the house down. When he was at Cambridge University he wrote down a list of sins, things that he regretted having done, and many, many years after the event he remembered having wanted to burn down the house with his mother and stepfather in it.

The science writer John Gribbin, Visiting Fellow in Astronomy at Sussex University, adds a little more about Newton's childhood. The more we discover about it, the more there appears a certain inevitability about the future course, if not the genius, of the man.

> He was a good practical experimenter, he built telescopes and so on, and this goes right back to his childhood when he would make really sophisticated toys. It is another example of the solitary child. His particular thing, which people wrote about years afterwards when he was famous, was that he made a working windmill which was driven by mice running round a treadmill. It amazed his contemporaries. The other thing was kites: he flew kites with lanterns attached to them and caused one of what must have been the first UFO scares in England as people saw these lights in the sky, down the pub afterwards!

Robert Iliffe has found out something of Newton's education and discovered, in an age of primitive education, a local system that would be the envy of many of the best schools today.

> We know that he went to Grantham Grammar School in his teens. He had as headmaster a man called Henry Stokes, who raised the level of knowledge of his students to that of a university.

Newton was already very well trained in mathematics before he went to Cambridge. He was probably much more mathematically educated than people used to believe, by the time he arrived there in 1661. Luckily enough we have a number of his notebooks and, in one of those, there are some notes on Aristotle. One gets the impression from reading them that he took those notes and then very quickly became bored with them and moved to the new philosophers then in vogue – Robert Boyle, Robert Hooke, Thomas Hobbes, Walter Charleton – and there you see this man questioning the whole tradition around him, as a young man, still an undergraduate, questioning what the state-of-the-art philosophers are doing. Yet he was not regarded as an outstanding scholar at first. When he went to Trinity College in June 1661 the person who took over his tutorial duties, Benjamin Pulain, had a number of other students and almost certainly left Newton to his own devices.

One of the great mysteries is his mother, who had a number of sheep left to her by her first husband and then came into a great deal of wealth by her second husband. She was extremely rich, earning something in the region of £700–£900 a year, which is about the level which a knight or a baron would have been expected to earn. Yet Newton went to Trinity College as a Sizar, as a sub-Sizar, a very poor scholar, someone who was forced to wait upon the

> other students and indeed, the Fellows, and was left to eat the remains of the food on the table after the others had left.

Newton came back to Woolsthorpe when the plague in 1665 forced him and others to leave Cambridge. Today in his old room there is his death mask, a telescope and a copy of the *Principia Mathematica* – Newton's greatest and most famous work. There is also a reconstruction of the prism experiment in a little cubicle in the corner of the room, an experiment he carried out there as a young man. Robert Iliffe explained its significance to me.

> We know that he was questioning the premier natural philosophers of the day. He was really doing some extraordinary experiments. He was very interested in colour. Maybe it was the case, as he later recalled, that he visited Stourbridge Fair, and there he bought a prism. He put a beam of white light through it and looked at the colours, the pretty phenomena that one could see when a white light was put through a prism and the colours placed on the wall. In addition to this he was doing a series of experiments on himself that we would consider almost barbaric, including looking at the kind of pictures that were made in his eye. He put various implements, such as a bodkin, which is an ivory toothpick, a brass plate, or indeed his finger, underneath his eyeball almost right to the back of

> the socket to change the size of his eyeball and hence produce colours. He wrote at the time:
>
> *I push a bodkin betwixt my eye and the bone as near to the backside of my eye as I can and pressing my eye with the end of it there appear several white, dark and coloured circles, which circles are plainest when I continue to rub my eye with the point of the bodkin.*
>
> He calmly notes down that after one of these experiments he was forced to lie in bed for two weeks with the curtains drawn.

I went with Robert Iliffe into Newton's study, from where we looked out at the remains of the great apple tree in what had been a massive orchard. The idea of thinking about gravity when an apple fell has everything. It is so simple it seems like something out of an ancient myth and it is something with which every child can instantly identify. Unlike Lewis Wolpert, who objected strongly to the myths about Archimedes and the bath, Robert Iliffe believes the story is important for what it represents, not as fact or fiction.

> It does not really matter whether the story is true, although it probably is. People are always looking for origin stories and it is lucky that Newton himself provided the most significant origin story of universal

gravitation that there is, and it is the apple from a tree in his own garden.

Newton called those two years he spent at home during the Plague 'the prime of my age for invention'. When he returned to Cambridge, he became the Lucasian Professor of Mathematics at the age of just twenty-seven. It is the post held today by Stephen Hawking. Of all the early scientists in this series, Newton is probably the one I would most have wanted to meet, although he was a solitary, an obsessive, a rather dislikeable man by all accounts. His enemies included the Astronomer Royal, John Flamsteed, the great German philosopher, Gottfried Leibnitz, and the physicist Robert Hooke. John Gribbin believes much of the fault lies with Newton himself.

He was very reclusive, he was a strange kind of person. He had this bizarre upbringing and he was a very solitary child. He hated publicity. When he was first becoming known as a scientist he loathed the response he got from what you would now call 'groupies' I suppose, people writing in asking him to comment on their theories and things like that. He hated the waste of time involved.

He had this huge run-in with Hooke about optics very early on in his career, when Hooke claimed that Newton's ideas about the way light works either were not original, or Hooke had done it all before or they were wrong. That really is what made Newton retreat into his shell and say 'I am not going to publish anything'. And although he did come out of his shell and publish the work on gravity and mechanics and so on, he waited until Hooke had died before he published his book on optics. He had it written, virtually the same, twenty years before. But he waited so that he could have the last word.

In my view he was downright weird. He hardly made any friends, he was obsessed by his work. There are many anecdotes about how on the rare occasions he had guests into his rooms in Cambridge he would go off to get a bottle of wine from the other room and forget why he had gone there and sit down and start working at his desk and hours later his friends would get up and tiptoe out and leave him in peace. It is impossible to psychoanalyse someone three centuries after the event and say what they were like and why they were driven the way they were, but one of the factors in his later life is that he became an unorthodox kind of Christian. He didn't believe in the Holy Trinity, which was a very dangerous thing not to believe in at that time, and specifically he would have been barred from holding office at Cambridge University by having those kind of unorthodox views. He did deeply believe in his own version of religion; he studied obsessively in the Bible and other sources. So there were sound reasons why he would want to keep relatively quiet during

> his time in Cambridge. We do not know really why, but he was very, very odd.

Despite the arguments and unpleasantness, most of Newton's contemporaries recognised that this was a genius living amongst them. John Gribbin asserts that even more important than his individual discoveries was the way in which he approached science.

> The really important thing about all of Newton's work is not what he did but the way he did it and the key thing is that he invented what is now the scientific method, the idea of actually doing experiments to test your theories and hypotheses. Before him the only person who had done that was Galileo and he never formulated it properly as a philosophy. Before Galileo, people used to go around following the tradition of the ancient Greeks of thinking beautiful thoughts and deciding different things like weights of different mass will fall at different speeds without ever bothering to test it. Newton actually wrote down that no idea is any good unless it has been tested and proved by experiment, and that is the key thing. Then from that he developed a series of laws, Newton's Laws, and the theory of gravity which applied to the whole universe. The other key insight from Newton is that the laws we work out with our experiments down here on earth are universal, literally, so we can tell

> about everything in the universe from these simple laws. That was a completely mind-blowing concept in the seventeenth century and it still is, really.

Sir Martin Rees, Astronomer Royal and Royal Society Research Professor at Cambridge University, takes up this point.

> Newton showed the power of mathematics in understanding the world. He was able to do this because astronomy was in fact the first science where exact measurements could be made. Astronomers had made measurements of time, and of the positions of planets in their courses across the sky for centuries before Newton, so there was lots of data. At that period, of course, there was very little exact data in any other science and that is one reason why it was astronomy that was the first science that lent itself to exact mathematical description. What Newton managed to do was to subsume all that was known about the orbits of the planets and the moon in a single law of nature where the force between any two objects depends on how heavy they are and on the inverse square of the distance between them, the so-called Inverse Square Law. He was able to show that this mathematical description actually allowed you to predict the positions of the planets in future almost like clockwork, and also to show that the law of nature which held the planets in their courses was

the same as the law of gravity that held us down on the earth and made the apple fall.

I suppose all sciences aspire to the ability to unify as much as possible in terms of simple laws that you can write down on a single sheet of paper, as it were, and which describe phenomena exactly, and Newton was the first scientist who actually achieved something in that direction. And the forces of the micro world, the forces that govern molecules and determine chemical behaviour, those took two hundred years before they yielded to a similar analysis – that had to wait almost to the twentieth century. So Newton was really far in advance of his time in applying mathematics to get precise results about the natural world.

Newton's most famous work is the *Principia Mathematica* – written in Latin and published in 1687. It is a massive achievement, perhaps the single most important book in the history of science. When he was asked how he had come to make the astonishing discoveries contained within its five hundred odd pages, Newton replied 'by thinking on it continually'. John Gribbin believes the *Principia*'s importance in the history of science cannot be exaggerated.

There are three laws of Newton's Laws of Motion that we all learn in school and the theory of gravity – all within the '*Principia*'. They are the laws which tell us the way things move, the way things bounce off each other, the way they respond to being pushed and pulled by forces and that underpin the whole of mechanics. They underpin everything from building the Forth Bridge so that it will not fall down when a train goes across it to sending a spaceship to Jupiter and, of course, if you are sending a spaceship to Jupiter, you also need to understand the law of gravity.

So it is really everything in the world that is mechanical, that moves around, that is described by Newton's laws, and it is those laws which people use to build the everyday things in the world, even things as mundane as washing-machines. The rate at which the spinning bits of the washing-machine spin round depends on Newton's laws and you use those laws in calculating the stresses involved. It is impossible to exaggerate its importance.

Before Newton, people still had this idea that the world was somehow capricious, that it was ruled by the gods and they might decide on a whim to make a different rule on Thursday from on Wednesday, something like that. But Newton says that if you are rolling a ball down a plane or if you are firing a cannon ball on a trajectory or even if you are working out how a sailing ship moves in accordance with the wind and the waves and the tides, all of these things come back to these very basic rules.

The first one is that things keep moving in a straight line unless something pushes them or pulls them. This he applied to the orbit of the moon. The moon tries to go in a straight line. Gravity stops it. The second is the idea that action and reaction are equal and opposite. Things bang together, they bounce off with equal force, and this applies with rockets, for example. The way you make a rocket move is by throwing things out backwards and that pushes you forwards. The third law tells you how fast things move when you apply a force. So what he is doing is putting order into the universe, saying 'Look, there are not mysterious things we cannot understand going on, it is not the whims of the gods just deciding how things ought to be, there are very simple, very basic things going on'. This leads to this whole idea now, three centuries later, when we think we understand how the universe was born and the Big Bang and things like that. It can all be traced back to Newton's insight that simple laws explain complicated things.

I asked John Gribbin if he thought Newton had stood on the shoulders of the giant Galileo.

He inherited ideas from Galileo, but I think Newton was such a special person that probably if Galileo had not existed, Newton would have done it anyway. But Galileo came up with some of these ideas. He nearly had the idea of inertia right. He thought natural motion was circular, which was not completely wrong, because he realised that something rolling on the surface of the earth would carry on because, of course, there is no friction and the earth is round, and Galileo thought that meant that circular motion was the natural way to move.

Some of Galileo's ideas were interpreted by Descartes and Descartes was very influential on the way that Newton thought. His famous quotation about the shoulders of giants did not apply to gravity, it applied to his work in optics. He knew there were great minds that had come before, but he was extremely special and I think he could have done it all himself.

How, I wondered, did he work out his law of gravity when he had no sophisticated equipment to test it?

This is an example of the way that he thought about things in universal terms and the way he broadened the scope of science by taking it off the surface of the earth and out into space. The story is that it was during the plague years, when he was back home away from Cambridge keeping away from the danger of the plague, that he sat around thinking about things like the orbit of the moon around the earth and the fall of an apple from a tree. And the key element of this was that he realised that they could be explained by the same law. The same thing pulls an apple off a tree and holds the moon in its orbit.

> Then he did a very simple calculation, working out the relative distance of the apple from the centre of the earth and the moon from the centre of the earth, and found that you could have the same law but make the moon fall at just the right speed to stay in its orbit. What happens is that the moon is sort of trying to fly off in a straight line, flying off at a tangent, and every second it falls a little bit towards the earth. That bends its straight line into a little curve, and to make the apple and the moon both fall in accordance with the same law it has to be an inverse square law. Then much later Halley came along and said 'Look, there is this row going on at the Royal Society about the orbits of the planets and their being ellipses, and how can you explain that?' And Newton said 'An inverse square law', so Halley says 'How do you know?' and he says 'I have worked it out somewhere, I have got it on a piece of paper'. Then he sent Halley away without the piece of paper because he wanted to double check everything and work it out properly, which he did.

I was intrigued to know how Newton could have been so sure he was right.

> Where Newton was different from his immediate predecessors, people at the Royal Society and his contemporaries, was that they had worked out that if you had an elliptical orbit it would match an inverse

> square law of gravity, but what Newton did was to put the mathematics in to prove that if you had an inverse square law you must have elliptical orbits. So before him people would have said 'Well, every orbit we know about matches an inverse square law, but we do not know that if we discover a new planet tomorrow it might not obey a different law, it might not match what we already know'. Newton said 'No. Here is a set of rules, equations, which tell you the way the universe works and any planet you ever find orbiting any star or any moon orbiting any planet must obey an inverse square law'. That is the difference. He was working from principles to general laws, not just taking an *ad hoc* view of what was going on and saying 'Oh well, that is what it looks like so far'.

The *Principia* was understood by very few of Newton's contemporaries according to Robert Iliffe, who quotes one bewildered early reader: 'There goes a man that hath writ a book that neither he nor anyone else understands'. Famously, one aristocrat offered £500 to anyone who could explain what it meant. Newton himself said he did not want 'smatterers' in mathematics bothering him with their questions. But despite this, through supporters in universities – often inserted there by Newton himself – Newtonian philosophy had a huge impact on seventeenth-century society, according to John Mullan, Lecturer in English at University College, London.

People did not feel in the eighteenth century that they should have read the *Principia*. There was a widespread acceptance that it was unintelligible to all but a very few people. I think it is telling that it was not translated into English until three years after Newton's death because there was no particular demand for the actual text. But there was a great deal of demand for popular lectures, lectures which often took place in coffee houses, and the important thing about that is that for the first time natural philosophy is a polite, genteel pursuit and that genteel audience includes women as well as men.

Accompanying what we now know was a sort of cottage industry of lecturing and demonstration, there is also the publication of popularising texts with titles like 'Sir Isaac Newton's Philosophy Explained for the Ladies' – that's the most famous title of all. But there are other similar texts for 'Gentlemen and Ladies'. That is something entirely new, the idea that what we would call science is something that genteel people should be interested in, the idea that science is fashionable, even voguish. It is something that I think is quite familiar to us, but it was entirely new at the time.

Jonathan Swift satirised Newtonian philosophy in *Gulliver's Travels*; Alexander Pope wrote that famous couplet: '*Nature and Nature's laws lay hid in night/God said "Let Newton be", and all was light*'. And some of the greatest intellectuals of the age, such as the philosophers John Locke and David Hume, acknowledge their debt to Newton. He was the first celebrity scientist, as John Mullan elaborates.

> He is endlessly depicted in portraits and in medals. He appears constantly in poetry. Perhaps the most famous example at the time, although now less well known, is the poetry of James Thomson, now only known for his writing of the lyrics of *Rule Britannia*, who wrote one of the most highly regarded and widely admired poems of the century called 'The Seasons', which has an extraordinary delight in the actual jargon, as we might call it, of the natural philosophy of the time. And in it he introduces a lot of new words which now sound to us like poetic words but at the time were actually scientific. Words like 'refracted', 'effulgent', describing the operations of light in particular. He translates that scientific enthusiasm into a poetic enthusiasm. We now recognise the idea of a scientist as a celebrated and revered person, but the personal fame of the scientist – which is, in Newton's case, even stranger given that he was an intensely reclusive, secretive, paranoid, deeply unlikeable human being – that fame is a kind of measure of something new that is happening in the culture.

Yet Newton did not spend all his time on his laws of motion. Perhaps bored by mathematics, he turned to alchemy and spent nearly thirty years carrying out peculiar experiments in bubbling cauldrons in his laboratory. For a long time scientists preferred to keep quiet about his interest in this ancient science, which seemed strangely at odds with his brilliant work on gravity and motion, but Newton may well have thought that his experiments in this area were as important as anything he ever achieved in the *Principia*. Unfortunately, we will never know exactly what he was doing. Dr Robert Iliffe has tried to pin it down.

It is very difficult to understand on many occasions what Newton was doing. For example, at one point in 1684 he said 'Today I made Jupiter fly on his eagle' and we do not really know what that means. Newton's whole chemistry was based on the language of alchemy. Using a furnace which burned continuously, Newton developed a whole series of amalgams and elements which we simply cannot replicate today. It is impossible to know whether various impurities entered the chemical material and made Newton develop various things that we simply will never be able to uncover. Alchemy was purported to be the province of adepts, gifted people touched by God. On a number of occasions in the 1670s and 1680s, he uses alchemical terms and language to shed light on aspects of what we would now call his science.

It is also important to remember that Newton was a very religious man. He was reading in the book of nature which God had written. In a sense he saw himself as a priest of nature.

At the age of fifty Newton had a nervous breakdown. There are several theories as to the possible cause. Perhaps he had been poisoned by the chemicals from his experiments. Perhaps it was the strain of having to keep his religious beliefs so secret. It has even been suggested that he had a homosexual relationship with a young Swiss mathematician which abruptly came to an end. Whatever the real reason, Newton turned on many of his friends, including the diarist Samuel Pepys, to whom he wrote a very strange letter:

Sir,
I am extremely troubled at the embroilment I am in and have neither ate nor slept well this twelve month, nor have my former consistency of mind. I never designed to get anything by your interest nor by King James' favour, but am now sensible that I must withdraw from your acquaintance and see neither you nor the rest of my friends any more if I may but leave them quietly. I beg your pardon for saying I would see you again.

Within a year Newton and Pepys were corresponding again about the mathematics of a new lottery, obviously trying to forget this unfortunate incident. And while this may have seen an end to Newton's scientific achievements, he did not retire from public life. Newton was made Master of the Royal Mint in 1698 and became the scourge of counterfeiters everywhere, sending some of them to the gallows. He became President of the Royal Society and was knighted by Queen Anne. When he died in 1727, his coffin was carried by some of the greatest men in the land to Westminster Abbey. Robert Iliffe describes the dramatic transition.

> The man goes from being a recluse to the most public figure of all on the grounds that he feels he has done it all in terms of his science. And when you have done the *Principia Mathematica*, there is really nothing else to do.

Near the end of his life, Newton wrote:

> *I do not know what I may appear to the world, but to myself I seem to have been only like a boy playing on the sea-shore, and diverting myself in now and then finding a smoother pebble or a prettier shell than ordinary, whilst the great ocean of truth lay all undiscovered before me.*

Two hundred and fifty years after his death, we still recognise our debt to Newton. I asked Robert Iliffe if Newton's assertion that 'If I have seen further it is by standing on the shoulders of giants' was characteristically or uncharacteristically modest?

> Well, it is certainly true. It is a very odd statement, of course, and Newton churns it out at regular

> intervals. It is not particularly modest. He is saying 'If I have seen further'. Well, there is no doubt in Newton's mind that he has seen further, but one gets the impression that he is actually not just standing on the shoulders of giants but stamping on the shoulders of giants. That is the amount of respect he had for people in the past. Not very much.

I asked the Astronomer Royal, Sir Martin Rees, whose shoulders twentieth century scientists stand on today, or whether he believes science is about moments of individual genius.

> When we look back on the history of science it certainly seems that it has evolved by a straight progression but, of course, there have been very many blind alleys, so science evolved and still does in a boisterous and contentious way, and we only see in retrospect the ideas that proved fruitful. But I think that, despite the fact that the way science advances is determined by all kinds of social pressures, political factors, economic pressures and all the rest, nonetheless most scientists do feel that they are engaged in an enterprise which is leading in an objective way towards a truer understanding of nature. So it is certainly true that scientists as individuals impose their personality on the subjects. They influence what other scientists think and what ideas become accepted.

> But in the long run if their ideas survive scrutiny they become just part of the corpus of public knowledge, as it were, and their individuality fades. Even modest scientists, all of us, have the satisfaction of feeling that what we contribute will be a durable part of the scientific edifice even though it may lose its identity and people will not remember that it was we who did it. If we go back to the seventeenth century, there were, of course, far fewer active scientists and Newton is a towering figure. I think most scientists would probably say that Newton may well be the most outstanding scientific intellect of all time. They would say he is not just the greatest figure of his time but I think they would rank him as number one in the overall panoply of scientific intellects.

Albert Einstein never doubted that Newton was the greatest scientific genius of all time. He said 'Nature to him was an open book. He stands before us strong, certain and alone.' Yet the common consensus is that in the early twentieth century Einstein stood on Newton's shoulders and somehow proved him wrong. Sir Martin Rees wanted to correct that view.

> People often ask what is the relation of Einstein's relativity to Newton's theory. Does Einstein supersede Newton? Does Einstein show that Newton was wrong? I think that is not quite the way to

look at it because even though we now know that Newton's theory is not exactly right – there are small deviations from Newton's theory, even in the motion of planets, the motion of the planet Mercury is slightly different, and light does not behave exactly in the way Newton expected – Newton's laws are still good enough for almost all practical purposes. When spacecraft are sent to Saturn and Jupiter, they are programmed according to Newton's laws. We can calculate the tide, predict eclipses essentially by Newton's laws. They apply, provided that bodies are moving much slower than the speed of light, so they only need modification when gravity is extremely strong or when motions are very fast, almost as much as the speed of light. It is then that we need Einstein's theory of gravity, which is something which reduces when motions are slow and gravity is weak.

Einstein's theory has wider scope than Newton's and indeed is essential when we try to understand some phenomena in the universe where we know gravity is much stronger than it is here on earth, that is, it can be applied even in extreme places like black holes and the Big Bang where Newton's theory does not apply. Things that seem somewhat arbitrary to Newton seem natural, almost inevitable, once we look at things in Einstein's way. That is the sense in which Einstein provides a deeper insight into Newton.

John Gribbin argues that, without Newton, the history of science would have been utterly different and inevitably the poorer.

> Einstein definitely thought that Newton was the greatest scientist who had ever lived and I think the fundamental reason is the one that I alluded to earlier, because Newton came along and had to do it all from scratch. Everybody since Newton has at least had the scientific method – Newton's method, if you like. You know where to begin and you know that there are universal laws, so you do not have to go out and make that huge conceptual leap of deciding that there is a way to describe the whole universe and how to set about doing it. You know how to do it and you know where to begin. So even somebody like Einstein in the present century, they may be huge intellects and it may be that if they were put in Newton's place they might have been able to do it, but Newton is the only person who did do it. He was first and that is what makes him pre-eminent.

Antoine Lavoisier

(1743–1794)

1743 Born in Paris, France.

1768 Elected to the *Académie des Sciences* in Paris. Joins the *Ferme Générale* as tax collector.

1771 Marries Marie-Anne Pierrette Paulze.

1772 Writes secret letter to *Académie des Sciences* containing his first thoughts on phlogiston.

1774 Joseph Priestley visits Lavoisier in Paris and shows him how to prepare oxygen. Publishes *Opuscules physiques et chimiques*, his first book.

1783 Demonstrates that water is not an element but a compound.

1787 *Méthode de Nomenclature Chimique*, a new dictionary of chemistry published. Instigates the building of a wall around Paris.

1789 Publishes the *Traité Elémentaire de Chimie.* Establishes *Annales de chimie*, a journal devoted to the new chemistry.

1793 *Académie des Sciences* suppressed. Arrested in Paris.

1794 Guillotined on May 8th.

The Revolution Does Not Need Scientists

THE CHEMIST Antoine Lavoisier was guillotined during the French Revolution at the height of the Terror in 1794. A contemporary declared 'It took them only an instant to cut off that head, but France may not produce another like it in a century'.

> *The declaration of the jury is that there constantly existed a conspiracy against the French people tending to favour by all possible means the success of the enemies of France: that Clément Delaâge, Louis Balthazar Dange Bagneux, Jacques Paulze, Antoine Laurent Lavoisier [and twenty-four others], ex-nobles and former farmers-general, are all convicted of being the authors of or accomplices in this conspiracy. After having heard the public prosecutor on the application of the law, the tribunal condemns the above-named to the death sentence.*

The dramatic story of Lavoisier's life, the stuff of novels of that period and of plays, is a tale of two revolutions at the end of the eighteenth century: in chemistry and in politics. Dr Simon Schaffer, Reader in History and Philosophy of Science at Cambridge University, brings the two together.

> Certainly many observers at the end of the eighteenth century think what they are living through is something like a revolution in the sciences and certainly a revolution in chemistry. Lavoisier himself describes what is happening as starting a revolution in chemistry and, what a lot of observers meant by a revolution, was at least partly that the sciences, chemistry in particular, were going back to their original principles, clearing away the errors of the ages and going back to simple, clear ideas from which the science could then be reconstructed afresh. I think then there is a very close link between that idea of going back to first principles and rebuilding the world of chemistry and, say, going back to first principles and rebuilding the human condition. That is why a lot of chemists were so closely involved both with the revolution in America and the revolution in France.

Yet Lavoisier is very far from the firebrand of humanitarian values which this implies. Unlike some of the other figures in the French Revolution – the passionate Danton springs to mind – Lavoisier comes across as a rather cold, aloof man. I spoke to his biographer Dr Jean-Pierre Poirier in the Lavoisier archives at the *Académie des Sciences* in Paris, where he is a member of the Comité Lavoisier.

> Lavoisier was a man of calculation; he had a very accurate, slightly obsessive mind. He was quite familiar with figures. I would say that he had a mind which was very logical, rather rigid. He was not such a bright person, but he was very logical.

Born in 1743 to wealthy parents, Lavoisier was a withdrawn child, preferring study to games. By the age of twenty-five he was a rising star at the prestigious *Académie des Sciences* in Paris. But unlike most of the other scientists in this book, science was Lavoisier's pastime, not his career. Realising that chemistry would never earn him a fortune, Lavoisier became a tax collector. He joined a private company called the *Ferme Générale* – General Farm – which collected taxes for the government and which together with the bribes and backhanders accompanying the position, gave him a very substantial income.

Professor Bernadette Bensaude-Vincent is Associate Professor at the University of Paris, Nanterre, and an authority on Lavoisier.

> The *Fermiers Générales* were hated by the people because they were in charge of collecting taxes and the taxes were very heavy, especially for poor people. Lavoisier himself was responsible for building

> a wall around Paris, again for collecting taxes, and this cost a lot of money. It also prevented the Parisian people from going outside Paris during the night. This wall caused hatred of Lavoisier.

Lavoisier married the daughter of his boss at the tax company, Jacques Paulze. Marie-Anne Pierrette Paulze was an heiress and only fourteen when she married the famous chemist. Independent-minded, even at such a young age, she had resisted the pressure to marry another suitor, a dissolute penniless aristocrat. Jean-Pierre Poirier comments on this eighteenth-century arrangement.

> The girl said 'I will not marry this gentleman' but the father understood that might be dangerous for his career, so he decided to marry the daughter as quickly as possible to somebody who would be acceptable. He would invite Lavoisier for dinner and there Lavoisier met Marie-Anne Pierrette. Lavoisier was perfectly acceptable. It was more a business arrangement than a love story. But, anyway, this couple was very happy for many years. Madame Lavoisier was a very bright person and she decided to have English and Latin and chemistry lessons and she became really her husband's assistant.

I asked Jean-Pierre Poirier if such an odd arrangement – by our lights – had produced a good marriage.

> Yes it did, at the beginning at least. They used to work together and, today, you can see drawings which were made by Madame Lavoisier. In them you see the entire team working in Lavoisier's laboratory.

Marie-Anne subsequently grew tired of the periods her husband spent away from Paris and began a passionate liaison with a mutual friend of her husband's. Lavoisier, apparently, did not object.

According to Madame Lavoisier, her husband called the one day a week he devoted to science his 'day of happiness'. The rest of the time he rose at six, worked on his experiments for two hours before going out on his tax collecting duties. In the evening he returned home to spend another three hours at his furnace.

Among his most famous experiments was a two-day-long public demonstration in which he proved that water was made of two elements, hydrogen and oxygen. He also renamed chemical compounds in his *Elementary Treatise on Chemistry*, which has been credited with having the impact on chemistry that Newton's *Principia* had on mathematics. Peter Atkins, Professor of Chemistry at the University of Oxford and Fellow of Lincoln College, is in no doubt about Lavoisier's enormous achievements.

> He really did two things. He drew the distinction between elements and compounds, so that people understood the way that the world was built much more clearly; and he found a way of attaching

numbers to chemistry. Quite an extraordinary thing really to attach numbers, measurements and so on, to matter. That was an amazing intellectual step in chemistry. It meant, to be more precise, that he used the chemical balance to weigh things and, as soon as you attach numbers to anything, you turn science into physical science and that gives you an enormous predictive power to investigate your ideas very precisely.

Lavoisier also overturned the theory of phlogiston, for which he seems best remembered. What, I asked Peter Atkins, was phlogiston supposed to have been?

I think to understand phlogiston it is easier to see where we stand at the moment in our understanding. If you think of setting light to a piece of magnesium ribbon then it burns brightly and forms an ash. We know now that what has happened is that the metal has combined with the oxygen of the air to give a compound, magnesium oxide. Before that was established, people thought that the metal, magnesium, was not a fundamental substance; that when you burned it, it released phlogiston as it was called, really from the Greek word for flame. It released phlogiston into the air and you were left with ash, which was the fundamental substance. So our modern view of combustion is quite the opposite of the theory of phlogiston. To overthrow the old

> theory, Lavoisier used measurements to show that materials became heavier when they burned in air. A clear way of doing this was to weigh the subject before and after it was burned and see whether it had lost weight – in which case phlogiston would be the right theory – or it had gained weight, in which case it would have been the combination with oxygen as the correct theory. Lavoisier's use of the balance established the latter.

I wondered that it had held sway for so long. Was it seen as something rather magical and mysterious, this phlogiston?

> I think people needed to make very precise measurements. In many of the experiments that people were doing when they burned things, things splashed everywhere, so it was not at all clear, even if they had weighed what they had got, that it had gained weight or lost weight. It was just too casual. Often people were using sort of magnifying glasses to get really high temperatures, by focusing the sun's rays on things, and they just dripped and ran everywhere. It was only when Lavoisier with, I suppose, his accountant's mind came along and really kept meticulous track of everything, that people were able to show that there really was a gain in weight.

Chemistry seems to have lagged behind physics. Newton

was working away defining the laws of physics more than a century before chemistry came into the modern age. I asked Peter Atkins if there was any single explanation for that.

> I think physics is concerned with primary qualities whereas chemistry is concerned with secondary qualities largely. Primary qualities are mass, length, speed and so on; secondary qualities are colour, smell, touch, and it is much more difficult to attach numbers to things like colour, things like smell. Chemistry moved much more slowly initially.

I wondered whether Lavoisier would recognise chemistry today, and what the key differences were between his and Peter Atkins' methods.

> Lavoisier really heated and stirred and splashed blindly and we today, when we look at our splashings and stirrings and heatings, see in our mind's eye the changes in the positions of the atoms that we are stirring around. So he stirred irrationally but hopefully; we stir rationally.

Lavoisier has often been accorded the title – given him most emphatically of all by the French – of the father of modern chemistry. But does he deserve it? Peter Atkins believes so.

> I suppose in a way he is really the executioner of alchemy. He swept aside all the awful names that

> none of us can remember, like tinctures of this and tinctures of that, which were simply based on the natural origin of things, like flowers and trees and so on which bore no relation to the composition of the materials. So he brought in an entire new language. He established the fundamental distinction between elements and compounds, which in a way the Greeks had. The Greeks had a wonderfully clear-sighted view of nature. They got things right in principle but didn't get them right in practice. They realised that there must be fundamental entities, the elements; they just got the identity of the elements wrong. Lavoisier helped to establish the correct identities of the elements and to distinguish them from combinations of the elements – what we now call compounds and so on. And third, Lavoisier introduced the art of precise measurement and so turned chemistry into a physical science. So, yes, I think he is properly regarded as the father of modern chemistry.

Yet there is no unanimity on this point. The historian Simon Schaffer's view is one echoed throughout this book.

> I have never been convinced that paternity suits are an interesting way of studying the history of the sciences. I would always want to emphasise the collective and shared and communal nature of scientific work through the ages.

Rather like Newton, Lavoisier was not always keen to credit others working in the field. Joseph Priestley was a brilliant English chemist and political radical, who visited Lavoisier in Paris and showed his French colleague how to prepare a new gas five times purer than ordinary air. Priestley called this 'dephlogisticated air', preferring to stick to the old theory. Lavoisier decided to call the new gas 'oxygen' or acid producer. But when he described his experiment to the French Academy he failed to mention either the work of Priestley or that of any of the other chemists with whom he had been corresponding. Simon Schaffer addresses this famous dispute.

> I do not think any of the protagonists of the chemical revolution were thieves. This makes it a rather exceptional event in the history of science. However, there is an enormous amount of unclarity about who is communicating which ideas to whom, and it is possible in retrospect to analyse that controversy as an example of plagiarism. Victorian hagiographers, both of Priestley and of Lavoisier, enjoyed enormously claiming that their hero had been done down and that the foreign enemy had misbehaved.
>
> I think that the image of the scientific genius has done more harm than good and indeed Priestley himself was extremely hostile to individualism in the sciences because he reckoned that the impression that the sciences are due to individual heroic discoveries has a catastrophic effect on the cultivation of the

> sciences right through society. And I think that is a healthier image, both then and perhaps even more importantly now, when we want to think about how the sciences are and should be cultivated.

Nevertheless, Priestley was extremely annoyed with Lavoisier for not acknowledging that he, Priestley, had played a part in his experiments and by Lavoisier's claim to have invented modern chemistry. Simon Schaffer sets the controversy in its wider context.

> It is no coincidence that we have a whole generation of French and British and other chemists who were working on very similar problems at the same time. We need to be thinking about the scientific community at the end of the late eighteenth century as extremely international, with very efficient means of communication: journals, learned societies, very quick correspondence networks, and also for the first time the beginnings of professional training.
>
> Both in Paris and certainly in the academies of Great Britain, we have laboratory training, people with shared techniques, shared experiences and, of course, shared equipment. The most fundamental aspect of this, and certainly something we usually forget, is the beginning of an international market in hardware. There is a very small number of London instrument-makers, glassware, balances, thermometers, and in Europe and world-wide.

To standardise equipment means to standardise problems. Along with the equipment you buy you get users' manuals, books, which train users and in that sense also standardise the questions the users ask.

In the dispute about oxygen between Priestley and Lavoisier, and I think this is often true in scientific controversies like this, there are right reasons and wrong reasons on both sides. Lavoisier is absolutely right that there is a substance like oxygen and that we find that in this particular part of the atmosphere, and he is right too that he has identified the crucial component of the atmosphere. He calls this substance, however, 'oxygen gas'.

Oxygen gas, according to Lavoisier, is a compound of two elements, which is not what we now think: oxygen, which is the principle of acidity, and caloric, which is the principle of heat. So in that sense Priestley and his successors are right to find problems in Lavoisier's doctrine. They are, for example, right to deny that all acids have oxygen in them and they are right to be pretty sceptical that there is an element – caloric – which is responsible for heat. So in retrospect this is a controversy in which everyone wins. Lavoisier clearly won, he developed a very powerful oxygen theory of chemistry. Priestley, ironically, in the end wins to a certain extent, through the next generation of British chemists and natural philosophers, who show that there is much wrong with Lavoisier's chemistry, notably his theory of acids and his theory of heat. Finally, and perhaps most importantly, Priestley was extremely sceptical of the kind of academic elitism which he associated with a lot of the work that Lavoisier and his French colleagues imposed.

In a very telling letter of the 1790s, after, in fact, Lavoisier had been guillotined, Priestley wrote to the French chemists saying: surely you do not want your regime to resemble that of Robespierre? Surely you do not want tyrannically to dictate what we should all believe? Surely you would rather win by persuasion than by dogmatism? He saw, roughly speaking, the wrong kind of scientific policy being used by the French chemists, as though one could only contribute to chemistry if one used their terminology. I find those positions relatively sympathetic.

I turned to Bernadette Bensaude-Vincent for a French opinion on the matter.

It was extremely competitive, of course, but I would say that, in the beginning, Lavoisier was very careful to acknowledge others' contributions. He is now considered as the discoverer of oxygen, which is not true really. He was the one who named oxygen, but he was not its discoverer. Lavoisier discovered nothing, in fact, in chemistry – I mean no new

> substance – in contrast to Priestley. But when he became confident, he became more and more proud of himself and he did not care to acknowledge the other's contribution, and he presented himself as the unique founder of the new oxygen theory.

Priestley's importance is often underplayed. He was a remarkable figure; his radical politics and his support of the French Revolution made him deeply unpopular in England. He is a fine example of the span of the Revolutionary impulse – from the laboratory to the Assembly or Parliament. I asked Simon Schaffer to summarise Priestley's significance.

> Priestley's role in chemistry is central because he co-ordinates a network of physicians, chemists, natural philosophers and industrialists, centrally concerned with the application of chemical understanding to industrial production. Priestley's allies include Josiah Wedgwood, the great potter and entrepreneur, for whom a knowledge of chemistry for glazes is absolutely central to commercial success; other industrialists like James Watt and Matthew Boulton, and medical reformers like Thomas Percival use Priestley's medical chemistry to rebuild and redesign hospitals.
>
> A better management of air, they agree, would be a better way of managing the social order. They use the term 'virtue' in two senses. Virtue refers to the quality of air, to support human and animal

> life, but it also has a moral sense. A good society is heavily correlated with a good natural order, of free circulation, free trade, liberal politics and almost literally an open-air environment. Priestley designed a series of techniques to measure air quality and these techniques were taken up in Italy and France to help reorganise the layout of markets, theatres, city planning, drainage and so on. I think we can speak of almost a Europe-wide campaign in aerial chemistry designed to produce enlightenment and reform, and it is in that context that Lavoisier and his contemporaries in Paris both learn a great deal about techniques and also begin to criticise them.

Whatever his rivalry with Priestley, Lavoisier was to discover far more dangerous opponents nearer to home. The revolutionaries, who had seized power several years earlier, had begun to turn their attention to the tax collectors. In the first years of the revolution, Lavoisier probably felt safe. He was himself a social reformer and had devised schemes for more equitable taxes, old age pensions and savings banks; his experimental farm at Blois was devoted to improving agriculture; and he headed work at the state Arsenal which had given the Revolutionary Army its first good supply of reliable gunpowder. Only when the revolution moved to the extremism of the Terror did his membership of the private tax-collecting consortium become an active threat. In 1793, the Sansculottes imposed an order seizing all papers belonging to the former *Fermiers Générales*. All they could find at Lavoisier's home were some letters in English from British scientists. These were taken. But at this moment Lavoisier seemed to care nothing more than that he might lose his fortune. His fame as a scientist, he thought, would save him. He made plans for establishing a new career as a pharmacist and wrote to the revolutionary committee in the third person, explaining that what he had done had been a long time ago and in any case he was now an important member of the National Committee on Weights and Measures set up under the new government. He wrote:

> *It is well known that he was never involved in the general affairs of the* Ferme, *which was conducted by a small committee appointed by the Minister of Finance. And, besides, his published works attest to the fact that he has always been principally engaged in scientific pursuits. He does not belong to the group of commissioners who were named to execute the decree regarding the rendering of the* Ferme's *accounts. Therefore he cannot be held responsible for the delay for which these commissioners are reproached. Hence he does not believe that he can be included in the law that ordered the* Fermiers-Générales *to be placed under arrest.*

At the time, the *Académie des Sciences* was housed in a section of the Louvre, which was considered a symbol of the privilege and elitism of the hated Ancien Regime and was closed down during Robespierre's Terror. Lavoisier, thinking that it was the last place that anyone would look for him, spent his last four days of freedom there, holed up in the empty chambers contemplating his fate. He walked out of it for the last time to give himself up to the Peoples' Tribunal. He was undoubtedly a revolutionary scientist, but would this be enough to satisfy the fanatics now in charge of the political revolution sweeping through France? Alas no. He was to spend six months in prison, with colleagues, including his father-in-law, pending trial.

It has been a subject of much speculation since then as to why his friends, and particularly his fellow-scientists, did not help him more and why he did not manage to escape the guillotine. In Jean-Pierre Poirier's opinion

> The main reason is probably the fact that most of his friends and colleagues were afraid to do anything because really, in those days, you could not say hello to a friend if this friend was a suspect because that could lead you to prison and to the guillotine. So I would say that fear was the main reason. There might be another one. Some of his disciples were slightly irritated by the fact that he was claiming that he was the only discoverer of what he had done and he was always a little reluctant to share with his colleagues and disciples a little of the glory that he thought was his glory. So maybe they were feeling frustrated by that behaviour. And the last reason is that he was such a rich man that people felt that he must be responsible for something, he could not be completely innocent.

Some of Lavoisier's friends did begin to realise the seriousness of his situation. One entreated Madame Lavoisier to visit the revolutionary Deputy who would decide Lavoisier's fate. She did so, reluctantly, but some say her arrogant manner only made matters worse.

In Lavoisier's last letter to his wife he shows none of the desperation he must have felt.

> *My dear one,*
> *You are giving yourself a lot of trouble, exhausting yourself both physically and emotionally, and alas, I cannot share your burden. Do be careful that your health is not affected. That would be the greatest of misfortunes. I have had a long and successful career, and have enjoyed a happy existence ever since I can remember. You have contributed and continue to contribute to that happiness every day by the signs of affection you show me. I shall leave behind me memories of esteem and consideration. Thus, my task is accomplished. But you, on the other hand, still have a long life ahead of you. Do not jeopardise it. I thought I noticed yesterday that you were sad. Why be so since I am resigned to everything? . . . If you have the chance to send a few bottles of table wine, it would be a great help to your papa who until now has been footing the bill for all the wine.*

According to Jean-Pierre Poirier, Lavoisier faced his death with courage.

Some witnesses say that he helped his companions

> behave. Some of them had decided to poison themselves, but Lavoisier said that they should not do that, because if they did poison themselves it would be some sort of recognition that they were guilty. The trial was on 8 May, 1794, in the morning, and they were condemned to death. A few hours after they were taken to the Place de Révolution, now the Place de la Concorde, and the twenty-eight of them were guillotined at five in the afternoon. But all the time Lavoisier kept a very dignified behaviour. His father-in-law Jacques Paulze was guillotined third of the group and Lavoisier fourth.

When Lavoisier was sentenced to death he is supposed to have pleaded for a couple more weeks to complete some scientific work. The judge's much-quoted reply was, allegedly, 'The revolution does not need scientists'. Lavoisier's scientific colleagues, like the chemist Fourcroy, did nothing to save him. Madame Lavoisier never spoke to any of them again. The greatest French scientist of the age went to the guillotine and his death is still a matter of embarrassment to the French, and still hotly debated among French historians, according to Bernadette Bensaude-Vincent. I wondered whether envy had played a part in this failure to come to his aid, or had it been fear of his supremacy, or his arrogance?

> I have no idea because Lavoisier's psychology is very difficult to understand. In my view, one crucial point

> is that the scientific ties did not resist the pressure of the political situation.

And yet only two years after his execution Lavoisier became a hero.

> Yes, it happened very quickly just after the Terror. A pompous ceremony was organised, a funeral ceremony too. A statue was unveiled and Fourcroy gave an eulogy. He presented Lavoisier as the hero of the chemical revolution and, although there was no connection between his scientific achievement and his end on the guillotine, he portrayed Lavoisier as a hero dying in the fight for truth, murdered by the brutish beast of the Terror, which embodied darkness fighting against the light of science.

This image of Lavoisier, the French hero fighting for true science, has remained a powerful one throughout history. I asked Peter Atkins whether he thought it was important for scientists to take into account the history of their subject and study the lives of their predecessors like Lavoisier.

> History is really for the aged. I think you need to be at least twenty-five or so before you start worrying about history. I think students really do not need to be introduced to the history of their subject except very casually. Modern explanations are so sharp and

> clear that you do not really want to clutter them up with the dust of ages. By the time you get old, into your mid-twenties, then you can sit back and start looking at science as a component of the cultural milieu and I think it is very rewarding to see the evolution of ideas and the trampling on the green shoots of ideas that has gone on through the ages, and then watch later on their final flowering. It is very exciting to look back over science and see how truth really has fought its way up through the rock formations in which it has been planted.

Simon Schaffer draws one final conclusion from the life and death of Antoine Lavoisier: that science and politics are inextricably linked. He argues that Lavoisier used the same skills and ideas to succeed in both his career as hated tax collector and as brilliant chemist.

> Lavoisier's execution was an extraordinary catastrophe, I think, for many of his colleagues and admirers internationally in the 1790s, because it seemed for many to symbolise the fact that the French Revolution had taken the wrong path and the slogan 'The Revolution has no need of scientists' was widely cited against what was going on in France.

Both France and England seem to have been rather unsympathetic to their chemists. The French chopped off the head of Lavoisier and the British more or less expelled the radical Joseph Priestley to America. His house in Birmingham was burned down by a mob in 1791. In Lavoiser's case, Simon Schaffer emphasises that the motives were purely political: that is, it was not chemistry but money – tax money – which led to his execution.

> The French did not execute Lavoisier because he was a chemist. They executed Lavoisier because taxation had been privatised and the people who run privatised tax systems tend to be treated relatively unsympathetically, especially in times of revolution. But I am not entirely convinced that Lavoisier was an unwilling political activist. In almost every sphere where Lavoisier and his allies worked, the balance is the central technique, the central instrument, and in many ways the central symbol of not just chemistry. You should use the balance in the sense of a mathematical balance, for example, to solve equations, they argued. You should use the balance in the sense of an economic balance to model how humans behave in markets and that is the essence of Lavoisier's chemical method, of his attitude to the human body and his attitude to society as well. Lavoisier says quite explicitly: the scientist in his laboratory can equally as the rhetorician in parliament contribute to the progress of humanity. And Lavoisier is clear that if we understand the grounds of human physiology, we will be able to organise society in a much better way.

In Lavoisier's last letter to his cousin, Augez de Villers, his coolness gets very close to real courage.

I have enjoyed a reasonably long and, above all, a happy life and I trust my passing will be remembered with some regret and perhaps some honour.

What more could I ask for? I will probably be spared the troubles of old age by the events in which I find myself embroiled. I shall die while in my prime, which I count as another of the advantages I have enjoyed. My only regret is not having done more for my family. I am sorry to have been stripped of everything and to be unable to give you and others tokens of affection and remembrance.

Evidently it is true that living according to the highest standards of society, rendering important services to one's country and devoting one's life to the advancement of the arts and human knowledge is not enough to preserve one from evil consequences and dying like a criminal.

I am writing today because tomorrow I may not be alive to do so and because I find it a comfort in these final moments to think of you and others who are dear to me. Be sure to tell those who are concerned about me that this letter is addressed to them all. It is probably the last I shall write.

Michael Faraday

(1791–1867)

1791 Born in Surrey, England.

1805 Becomes bookbinder's apprentice.

1813 Sir Humphry Davy employs Faraday as his assistant.

1821 Demonstrates electromagnetic rotation. Enters Sandemanian Church and marries Sarah Barnard, a fellow member of the Church.

1823 Liquefies chlorine.

1824 Elected to the Royal Society.

1825 Discovers benzene.

1826 Begins Christmas Lectures for children at the Royal Institution.

1827 Succeeds Sir Humphry Davy to Chair of Chemistry at the Royal Institution. Publishes *Chemical Manipulation*.

1831 Discovers electromagnetic induction.

1834 Publishes laws of electrolysis.

1837 Introduces ideas of lines of force.

1839–55 Publishes *Experimental Researches in Electricity*.

1844 Excluded from the Sandemanian Church in March but reinstated five weeks later.

1845 Shows that polarised light is rotated by magnetism.

1846 Lecture on ray vibrations which inspires James Clerk Maxwell's work on electromagnetic theory of light.

1849 Lecture on 'The Chemical History of a Candle'.

1859 Publishes *Experimental Researches in Chemistry and Physics*.

1867 Dies.

The Great Experimenter

MICHAEL FARADAY, born in 1791, is generally held to be one of the greatest scientists of all time, and not just by other scientists. One of Britain's few Prime Ministers with a scientific training, Margaret Thatcher, paid him this sincere but perhaps rather self-projecting tribute:

> He was a remarkable person, typical of so many of our great people, very ordinary background: father ran a smithy, son had virtually no education but he was brilliant. I admired his scientific discovery, I admired his method – you know, the experiment, the proof and the imagination. I admired his total fascination with the subject. His total singleness of purpose.

I can remember from chemistry lessons that Faraday discovered the laws of electrolysis and from physics classes I could also have told you that he was the first man to generate electricity from a magnet. But I had not realised that Faraday's later theoretical work radically challenged the ideas of physicists who had preceded him.

It is this sudden shifting of the perspective of knowledge that is so thrilling in science. Faraday was one such – a man who radically changed a way of looking at the world. He was so important to a subsequent generation of scientists that Einstein himself said that Faraday and his great successor James Clerk Maxwell had made the greatest breakthrough in physics since Isaac Newton.

Michael Faraday spent his life working at the Royal Institution of Great Britain in Albermarle Street, a beautiful building in the heart of London, which is still occupied by the Royal Institution and where the basement laboratories in which he carried out his experiments have been well preserved. His magnetic laboratory is small, rather bare, with much less equipment than a laboratory in a thousand and one modern schools, and it was here that Professor Sir John Meurig Thomas, Master of Peterhouse and Director of the Royal Institution until 1991, showed me Faraday's lines of force using iron filings and a magnet, now preserved behind glass.

Seeing the beautiful patterns created was every bit as moving as seeing a Shakespeare First Folio or even the Lindisfarne Gospels. And it is a true tribute to Faraday, I think, that this outward archive of the working of his mind should be so easily available to the public. In a basement in one of the richest areas of one of the richest cities in the world are what could be called the tools of trade of a man who changed that world.

If you take one of the blue seats in the great lecture hall you can look down on the kidney shaped table and see the exact spot where Faraday himself once stood to give his famous public lectures. It was in these imposing surroundings that John Meurig Thomas, a great enthusiast for Faraday, explained to me how the son of a poor blacksmith became the greatest scientist of his day.

He was a bookbinder. He started his working life delivering papers, actually. He left school at about thirteen. He could do arithmetic, reading and writing, but he was clearly an extremely curious individual. He had an enquiring mind. He started keeping a notebook about observations and thoughts about the nature of life and art and history and natural philosophy. He started doing that when he was about fifteen or sixteen. He was working in a bookbinder's shop and one of the customers there who was a member of the Royal Institution – this was in 1812 – gave him tickets to come and listen to the star of scientific Europe, Sir Humphry Davy.

Davy was a poet, a friend of Coleridge and many others. Davy was also a brilliant experimentalist. Faraday saw the experiments that he gave, dazzling even now: we repeat them here and it still brings the house down. He sat up in the gallery above the clock and listened, mesmerised, to this man giving one lecture after the other. He wrote up those lectures, bound them, sent them to Davy and

> that in effect was it. It's a long story, there were lots of hiccups on the way, but Davy hired him as a bottle-washer. Within six months he was doing some of the most intricate chemical experiments that you can imagine, preparing capriciously explosive materials. Davy had just got married, and took Faraday to Europe with his new wife. They travelled with their laboratory throughout Europe and he came back having been in a personal university, under the tuition and tutelage of Humphry Davy. That is how he came into science.

Michael Faraday's career first as a chemist and then as physicist owed much to Humphry Davy, but the relationship was a complicated one. The young Faraday rapidly proved he was worthy of Davy's attention. Working as Davy's assistant, in a sense swapping his bookbinder's apprenticeship to become an apprentice to science, he quickly began to carry out his own experiments. Iwan Morus, Lecturer and Wellcome University Award Fellow at Queen's University, Belfast, illustrates how Faraday's keen intelligence and brilliance as an experimenter rapidly led to bad feeling between the two.

> There are a couple of episodes in the 1820s when things got into quite a mess. For example in the mid-twenties, when Faraday was doing his work on chlorine. Davy initially was the one who told Faraday 'Michael, go and do this'. Then when

Faraday wrote up the experiments to be sent to *Philosophical Transactions* for the Royal Society, Davy insisted that Faraday put a footnote, so to speak, to the paper saying 'Sir Humphry Davy instructed me to carry out these experiments'. The relationship was always very complex and of course Davy's wife is an added ingredient in the mix. I mean it is well attested in Faraday's own correspondence. Lady Jane treats Faraday quite unambiguously as a servant and Faraday gets very uptight about the way he is being treated.

In a not untypical letter to his friend Benjamin Abbott, Faraday wrote:

Rome, January 25th, 1815.

Dear Benjamin,
I should have but little to complain of were I travelling with Sir Humphry alone or were Lady Davy like him, but her temper makes it often times go wrong with me, with herself and with Sir Humphry . . . She is haughty and proud to an excessive degree and delights in making her inferiors feel her power. At present I laugh at her whims which now seldom extend to me but at times a greater degree of ill humour than ordinary involves me in affray which on occasions creates a coolness between us all for two or three days.

After his work on chlorine, Faraday turned his attention to electricity. He was the first to show that a magnetic field could produce a current. This is why most science dictionaries credit him with the invention of a primitive electric motor. It made Faraday's name, but caused his relationship with Sir Humphry Davy to sour even further, as John Meurig Thomas explains.

> There is much misunderstanding about this. Davy was a vain man. He was a passionate man. He loved the applause of the multitudes far more than Faraday. What happened was that in 1821 – this was the turning point – Faraday heard another great scientist talking to Davy about the experiment that had been done in Copenhagen by another great scientist by the name of Oersted, a Danish medical scientist, who had spotted that if you passed a current through a wire then a needle of a compass would twitch. This was a significant observation. What did this mean? What should he do with it? Faraday went away and worked out how to create an electric motor from that idea. He sent the paper off, it was published and it brought him international fame. Davy was very annoyed because he thought that the idea had come really from him, and more especially from Wollaston.
>
> They both felt sorry afterwards, especially when Davy, who was then President of the Royal Society, started imputing in public that Faraday had stolen

this idea. It was all rather messy, but it was patched up. In fact Davy voted against Faraday for election to the Royal Society. It shows that he had a streak in him that was unfortunate. I have read a great deal of Davy's work, I would imagine he was deeply sorry for that. That is the sort of man he was. A very human individual. One does not have to belittle Davy in order to extol Faraday.

Faraday loved him right to the end. When Faraday was touching seventy, he wrote an account of his experiments, and he writes a footnote on the first ever paper that he published. He said 'This is a very precious paper for me. I published it as a result of work given to me by Humphry Davy at a time when my fear far exceeded my confidence and when both exceeded my knowledge'. He pays homage to Davy – there was no enmity, no permanent enmity, between them but there were periods of intense friction.

So this was to do with Davy's vanity?

Vanity and jealousy. I mean, he saw this man who was going to be better than him. Very hard to take. It really must be the same these days.

If it was hard for Faraday to build his career as a man of science then that was partly because at the time there was no such person as 'a scientist'. The word 'scientist' was not coined until the 1830s, and even then the position of 'scientist' was not respectable. The poet Coleridge was not the only one complaining about the middle-class men interested in science who were calling themselves 'natural philosophers'. So 'scientist' as opposed to 'philosopher' originally had a rather pejorative meaning: not a title for a gentleman. Faraday always preferred to think of himself as a natural philosopher, in the tradition of Newton. John Meurig Thomas elucidates this.

> You have to remember natural philosophy is the phrase that the Royal Society used to cover everything that we now regard as physics, chemistry, geology, et cetera. Natural history would deal with plants and so forth. I think at the time there was nothing really deeply profound in that. I mean he loved to call himself a philosopher, a lover of nature and natural knowledge. I think the phrase tripped off the tongue better. It also seemed to match his aspirations more precisely. I do not think there is anything more profound in it than that.

Whatever Faraday chose to call himself, he had to think very hard about how to build his own career as this new person – a man of science. The historian Iwan Morus spells out Faraday's difficulty.

> When you think about it, the kind of position

somebody like Faraday was in meant that there simply was no model. How does he figure out how it is that somebody who is going to be a natural philosopher is meant to act? I think very deliberately Faraday sets out to fashion himself in a particular way from almost as soon as he is employed at the Royal Institution. When he starts lecturing he asks Benjamin Smart, his elocution teacher, to come to the lectures so that he can listen to how Faraday comes across and correct his presentations. He would have friends literally hold up cards with 'too fast' or 'too slow' so that Faraday could tailor the way he performed. That is clear from any number of contemporaries' accounts of what it was like to be at one of Michael Faraday's lectures. Faraday clearly learned his lessons extremely well.

Here are two paragraphs from what became his lecture on the chemical history of a candle which he gave in 1849. Perhaps because of its graphic, even biblical, simplicity, it caught the imagination of the public.

Now I must take you to a very interesting part of our subject – to the relation between the combustion of a candle and that living kind of combustion that goes on within us. In every one of us there is a living process of combustion going on very similar to that of the candle, and I must try to make that plain to you. For it is not merely true in a poetical sense – the relation of man to a taper; and if you will follow I think I can make this clear.

. . .

You will be astonished when I tell you what this curious play of carbon amounts to. A candle will burn some four, five, six or seven hours. What then must be the daily amount of carbon going up into the air in the way of carbonic acid! What a quantity of carbon must go from each of us in respiration! What a wonderful change of carbon must take place under these circumstances of combustion or respiration! A man in twenty-four hours converts as much as seven ounces of carbon into carbonic acid: a milch cow will convert seventy ounces and a horse seventy-nine ounces, solely by the act of respiration. That is, the horse in twenty-four hours burns seventy-nine ounces of charcoal, or carbon, in his organs of respiration to supply his natural warmth in that time.

Faraday also made these Friday evening discourses dramatic occasions. One evening he placed water in two iron containers and left them in the basement to freeze. When they later exploded the audience was shocked, but Faraday had proved that water expands when frozen. His lectures because weekly events, and the whole of London flocked to the Royal Institution. John Meurig Thomas describes it as a great theatre of knowledge.

You had Charles Dickens occasionally, you had Charles Darwin frequently, you had Charles Wheatstone, a great scientist, invariably listening to him on Friday nights. He brought new discoveries to people's knowledge, but he was also in a quiet way a supreme showman. He believed it was his duty not only to show new discoveries but to explain them.

For example, let me tell you one of the incredible things that he did in this theatre. He built a cage, a cube, twelve feet by twelve feet by twelve feet, of wood. He covered it with metal, metal foil, metal wire. He stepped inside with delicate electrical instruments and then he had his assistant charge up this cage to something like a hundred thousand volts. All the people around were watching sparks flying everywhere, and he was inside not feeling a tremor. He predicted that. What a hell of a thing to do! You could be killed if you did not know your science well enough, and he came out and that was that. That is one side of his showmanship.

He had that impact, but then look at the magnitude of his discoveries. He liquefied about twenty different gases. Refrigeration becomes possible as a result of that. It did not have an immediate impact in that direction, but his laws of electrolysis changed the nature of industry and manufacture. When he demonstrated in one series of experiments in 1832 that the electricity of the clouds – frictional electricity – and animal electricity – the electricity of the electric eel shall we say – and the electricity of the dynamo were all the same, that was a very major step forward, which transformed the picture of experimental science.

But his biggest impact theoretically, which we are still feeling and which Einstein and Clerk Maxwell and all the great physicists since have applauded repeatedly, is that he worked out an idea which was absolutely right but nobody had thought about before then: the notion of the field. You see, what he really said, what that really means, is if you have a magnet the force of the magnet is not contained within the perimeter of the magnet – we all know that, every child knows that, but how far does it go? He argued that you have to think in terms of the natural world, not just in terms of Newtonian mechanics which enable you to predict eclipses and so forth, the ebb and flow of the tide and so on, you have to think of a field of a body exerting its magnetic or electrical or gravitational influence right throughout the end of space. The French physicists, natural scientists, had worked out the inverse square law. Faraday was not content with that. He wanted to know – there must be some influence of the medium in between, and this was a conceptual breakthrough.

> That is what has given rise to electronics – the fax machine, telephone, television, the wireless, the radio, the gramophone: they all go back to Faraday's understanding that you can have this force in the ether which you can tap and harness and pull out. So it is a magical thing. They all go back step by step to Faraday.

To a non-scientist, it is often the small things that snag the attention. For instance I find it ironic that Faraday was living at a time when mathematical physics was growing in importance and yet he could not do the simplest sums. Meurig Thomas took my question about this more seriously than it deserved.

> Many people think you cannot be a theoretician unless you are a very adept mathematician, and that is quite wrong. Just to give you another example: Charles Darwin has influenced man's view of the cosmos and man's position in the nature of things as much as anyone has. He did not have a simple equation, he had a theory, the theory of evolution. You could follow those theories. The incredible thing about Faraday is he discovers how to convert magnetism into electricity, he does so over a period of ten days spread over ten weeks. He sits down and thinks hard and comes to the conclusion of what you need and he postulates it in a theorem. He says if you cut lines of force you will generate

electricity. If you cut a large number of them, in
other words if there is a high density of magnetism,
you create more electricity. If you move very
quickly you create a greater impact. It was the
theory of electrical magnetic induction, expressed
in words, conceptually perfect. It has never been
modified. But he did not enunciate it in the form of
equations.

Now, Maxwell's approach was totally
different. He wrote it down in three everlasting
equation's – Maxwell's Equations. It has
been said that when mankind dies away and
even when Shakespeare's poetry is forgotten,
those Maxwell Equations will stay for ever.
They are some of the greatest equations that
you can have. But Faraday's method was not
mathematical.

James Clerk Maxwell, born in 1831, one of the most brilliant theoretical physicists of all time according to John Meurig Thomas, submitted for his Cambridge thesis work on Faraday's lines of force.

With that work in 1855, we understand that
the nature of electromagnetic radiation – all
the light that you can have from gamma rays,
from X-rays, right through to the infra-red,
to radio waves – they are all the same. They
have a unity and they will be understood

> by the principles initially perceived very vaguely by Faraday and fully understood by Maxwell.

John Meurig Thomas says that the extraordinarily fortunate coincidence of the careers of Faraday and Maxwell makes one 'believe in the Almighty or something'. Here is a letter written by Faraday to Maxwell:

> *November 13th, 1857.*
>
> *My dear Sir,*
>
> *There is one thing I would be glad to ask you. When a mathematician engaged in investigating physical actions and results has arrived at his own conclusions, may they not be expressed in common language, as fully, clearly and definitely as in mathematical formulae? If so, would it not be a great boon to such as we to express them so – translating them out of their hieroglyphics that we might work upon them by experiment? I think it must be so because I have always found that you could convey to me a perfectly clear idea of your conclusions, which, though they may give me no full understanding of the steps of your process, gave me the results neither above nor below the truth and so clear in character that I can think and work from them.*

> *If this be possible would it not be a good thing if mathematicians, writing on these subjects, were to give us their results in this popular useful working state as well as in that which is their own and proper to them?*
> *Ever, my dear Sir, most truly yours.*

It is a suggestion that has gathered increasing relevance over the last one hundred and forty years.

Faraday's ideas inspired his own century and ours. But he probably would not have liked the rather simple image we have of him as the inventor of the electric motor. It is rather more complicated than that, believes Iwan Morus, who disagrees with John Meurig Thomas that Faraday's work leads straightforwardly to twentieth-century technology.

> He is a massively important figure. Without Faraday I suppose we would not have the kinds of theories on electrical magnetism that James Clerk Maxwell puts together in the second half of the nineteenth century which lead on in a variety of ways to rather a large chunk of twentieth-century physics. So yes, Faraday is a massively influential figure but not necessarily influential in the way that various myths have grown up around Faraday tell us. Very little of what Faraday himself did had anything to do practically with the emergence of electrical technologies in the later nineteenth century, despite

> adverts now to the contrary. Faraday did not invent the electric motor. Faraday did not invent the electric light bulb. Faraday did not invent any particular technology and in fact Faraday himself would have been most horrified most probably at the imputation that he was a mere inventor. As far as Faraday is concerned, he is a discoverer of great natural philosophical principles. He is certainly not going to be engaged in the rather sordid business of inventing, which is something that craftsmen or entrepreneurs or people who are not gentlemen do.

If it seems strange to us at the end of the twentieth century that Faraday could rise from a poor and initially ill-educated background to be one of the greatest scientists ever, perhaps it is even harder to understand that this great scientist was also a strict believer in Biblical truth. Today many of us try to reconcile our belief in modern science and our need for faith, but for Michael Faraday there was no such conflict. He believed that his work, his science, was about understanding the world as God created it. Faraday was part of a small sect called the Sandemanians, which evolved after a schism in the Church of Scotland in the eighteenth century. His family were early members and Faraday joined the Church just before his thirtieth birthday.

Geoffrey Cantor, Professor in the History of Philosophy and Science at the University of Leeds, has made an important study of Faraday's religion and the relationship it had with his work as a scientist.

> First and most important is the commitment to live one's life by the Bible and in imitation of Jesus Christ. And this sort of commitment to a high degree of individual religious morality is really at the centre of Sandemanianism. Over and above that one immediately commits oneself to being a dissenter, one is not of the Church of England. For example, Faraday does not even go along to the memorial services of quite a number of his friends and their famous contemporaries because he feels as a Sandemanian he cannot participate in the Church of England. Over and above that, one is committing oneself to live in a very small and insular community. There is a high degree of intermarriage. For example, Faraday and his brother marry sisters. Their father, Edward Barnard, was an elder in the London Meeting House, the Goldsmith. This was very typical, two Faradays marrying two Barnards and this is another way by which Sandemanianism is maintained. It becomes very much a family religion. It is the duty of the community to provide for them. So you find Faraday, who at one moment might be hobnobbing with some of the key scientists of the day, later in the day going off to visit some elderly lady. This provides almost a dichotomy in Faraday's life, that on the one hand he is part of

> this small insular group and on the other hand he is relating to the larger scientific community.

John Meurig Thomas adds his observations on Faraday's religious life.

> He was driven largely to study nature because of his religious commitment. The Sandemanians were a very peculiar sect, but he believed. The most important thing in his life was his religion. What he was doing in looking at nature was to see the manifestation of the Almighty. This comes through in his brilliant articles. Let me just tell you there is one famous article, which he wrote in 1845, when he found some glass that he had made by a very special chemical means. When you passed a polarised ray of light through it and put a magnetic field on it, its polarisation was tilted. Now that is a hugely profound experiment. Magneto-optics began in the basement of this building on that day.
>
> When he writes it up he says '*I have long held an opinion almost to conviction in common, I believe, with other lovers of the natural worlds, that all the forces of all place matter have one common cause*.' This was his religious picture, you see.

Late in his life, remembering his early apprenticeship to a bookbinder, Faraday explained that science offered less conflict with this religious view. It had been a way out of what he called the vicious and selfish nature of trade. This seemed to me a rather idealised view of science and scientists – even for the nineteenth century. Geoffrey Cantor again brings it back to Faraday's religion.

> The Sandemanian community was a band of brothers and sisters, and indeed when they write to one another they head it 'Very dear brother', 'Very dear sister', and this was looked upon as a spiritual brotherhood. Faraday found the scientific community had some of these values and when he writes about the scientific community it is often very much in glowing terms. He talks about the scientific community being a band of brothers and a lot of his close scientific colleagues were people whom he related to very much as brothers, people who shared the same values, even if they were in a very different social or religious situation. But at the same time he viewed them as being committed to an ideal of pursuing science which was not trammelled by politics.

Faraday's own relationship with the Sandemanian Church was at times a difficult one. In the 1840s a major disagreement arose amongst members of the Church and, along with several other members, Faraday was excluded. It is not clear exactly what caused this schism, but undoubtedly it sent him into a deep depression and he had what we would probably call a breakdown. Although he was reinstated in the Church soon after, Faraday lived in deep fear of being excluded a second and permanent time. Yet, interestingly to me, Faraday does not seem to have faced any of the big questions about the relationship between science and religion which confronted some of his most important contemporaries. Geoffrey Cantor believes he deliberately avoided those areas of study he thought might bring him into conflict with his strict beliefs.

> Faraday very much tailors the kind of science he pursues. He had an early interest in geology and a lifetime interest in natural history and he would often go out in the countryside with a small botany book relating the names of plants to what he saw. He has very little to say about some of the new and challenging ideas. Although he lives for seven or eight years beyond the publication of Darwin's *Origin of Species*, he never makes any comment on this – true, he is ill for much of this period, but one also has some of the feeling from earlier comments on not dissimilar theories that he looked upon such views as speculation running well ahead of evidence, and he really does not pay them all that much attention. So in a strange sort of way his own science leaves many gaps and although he is familiar with Darwin and indeed with many of the people in biological sciences who are to move it in directions which do raise the problems of science

> and religion for many people, he really does not engage their theories.

Whatever the criticism of Faraday – that he did not take much interest in Darwin, or that he never himself gave a hand up to an apprentice as Davy had done for him – there is no doubt that he was an inspirational figure. He was also a famous one in his own lifetime: his correspondence includes a letter from Napoleon Bonaparte, the future Napoleon III of France. He was, as some scientists today are again becoming, an intellectual star. I do not think it is stressing his importance to us non-scientists too much to say that a large part of his legacy must be his dedication to popularising science.

Ravi Mirchandani is Associate Publisher at Weidenfeld and Nicolson and has published many of today's best-selling science authors.

> I think one of the things that was unique about Faraday (and there are very few people that you could put in that bracket) is that he was an absolutely major scientist and a very important part of the history of the popularisation of science. What he was lecturing on were issues much more akin to the kind of basic science that we are taught in schools, and the kind of basic science that puts an awful lot of people off science when they are at school. I think one of the reasons why so many people whose education has been in the arts or

humanity side read popular science today is because it addresses those big questions that their school science did not answer.

But that is not the kind of populariser that Faraday was. He was talking about the basic mechanics, as in his famous lectures on the candle, using a candle flame as the route to explore a whole variety of things about chemistry and physics, in a way much more akin to what somewhere like the Science Museum does, or the Exploratory in Bristol. I don't think there is a straightforward trajectory, though, since his time. There have been ups and downs. Probably the last time that science popularisation was a very significant area prior to now was in the Thirties and Forties involving both biologists and physicists at that time. Now, over the last ten or fifteen years, you have another upsurge of it. There are many people now whose popularisation of science, even if they are scientists and even if they have Chairs at universities, is actually more important than anything they will achieve as scientists.

John Meurig Thomas is also exercised on this topic of popularisation.

It is a great thing that a man like Steve Jones talks the way he does, in an extremely lucid and passionate manner and in a very communicable mode to the general public. He has the knack of actually picking up what it is that people would like to know. Not enough scientists have done that, which is a pity.

Of course many people have got very cynical about science. Medicine and chemical contributions to medicine and pharmacology and pharmaceuticals have been extremely profound and yet people will tend to concentrate on the evils of science and therefore want to turn away from science – those who are ignorant of it. What does depress me, I must confess, is that some cultured people whose poetry I like listening to, or whose architecture I like admiring, are even disdainful when talking about science. They say 'I know nothing about science and I'm proud of it'. I find that a bit sad, actually.

In Faraday's time things were rather different. He was both the greatest scientist of his day and an important public figure. His influence both on the scientific and the literary community of his day inspired one famous novelist to write with a suggestion for bringing his work to an even wider audience.

May 28th, 1850.

Dear Sir,

It has occurred to me that it would be extremely

beneficial to a large class of the public to have some account of your late lectures on the breakfast-table, and of those you addressed last year to children. I should be exceedingly glad to have some papers in reference to them published in my new enterprise, Household Words. *May I ask you whether it would be agreeable to you and if so whether you would favour me with the loan of your notes of those lectures for perusal.*

With great respect and esteem I am, Dear Sir, Your faithful servant,

Charles Dickens.

Faraday died at home in his study in 1867. His legacy to science and to the popularisation of science is indisputable. But the image I had of Faraday does not always tie in neatly with reality. He rose from an undistinguished background to be the greatest natural philosopher of his generation, at a time when natural philosophy was a gentleman's occupation and he was a blacksmith's son. He was deeply and strictly religious and spent most of his life with his fellow Sandemanians, yet he also loved the brotherhood of science and was equally happy in his basement laboratories at the Royal Institution. Perhaps these contradictions should only increase that admiration for his enormous achievements. I am happy to leave the last words to John Meurig Thomas, speaking in the Royal Institution where Faraday made such a mark on the world.

He went through his Bible sedulously, reading it

every day. He had a special code and special lines for marking it. His favourite verse was in the Book of Job: 'If I justify myself, my own mouth shall condemn me. If I say I am perfect it shall also make me perverse.' His religion taught him humility and never to be boosting your own position and to give the benefit of the doubt to the others. But there are some insights which Tyndall, his great admirer and successor here, has revealed. Tyndall says that Faraday had a fiery temperament. He had the rage of a volcano inside. But he had high principles and discipline, and instead of frittering away that passion on useless things and anger, he would turn it into a more positive attitude.

This is what is depressing about Faraday. Not only did he have technical virtuosity, he had moral perfection. You know, he was angelic. There was a lovely letter that Tyndall wrote to him in 1855, saying *I have just been to the British Association. It was very acrimonious, very distasteful – I disliked it intensely*. Faraday writes back: *My dear Tyndall, Listen to an old man*. (He was sixty-four.) *I have learned that when you find that people are aggressive and nasty towards you, be dull in your apprehension. Play it down or ignore it. But if they are trying to give you some praise, then be quick in your appreciation*.

He would work all this out, you see, in every conceivable way. There is not a thing that Faraday did which does not have a freshness about it. Let me tell you about the Friday night discourse in this lecture theatre since 1826. Nobody introduces anyone. When the clock strikes nine, those doors open and in comes the speaker. You are straight into the lecture. When it ends, the clock strikes ten, rapturous applause. He is led out by the Director. You are not introduced and you are not proposed a vote of thanks. A very good method for a lecture. It is theatrical. Perfect, you see. And Faraday thought that out and we still carry it on. And in many, many ways he brought his skill to human presentation and human activity. He was a man in a thousand million. You know, you ask what was the source of Faraday's genius. Well, it is the confluence of all those genes and the environment, but he was what he was. *Sui generis*, like Shakespeare. You cannot really analyse the genius of Shakespeare or Mozart likewise. I put Faraday in that class.

Charles Darwin

(1809–1882)

1809 Born in Shrewsbury, England.

1825 Enters Edinburgh University to study medicine.

1828 Enters Christ's College, Cambridge, intending to become a clergyman.

1831–6 The voyage of the *Beagle*.

1836 Made a fellow of the Geological Society.

1839 Publishes *Journal of Researches into the Geology and Natural History of the Various Countries Visited by H.M.S. Beagle*. Elected to the Royal Society. Marries Emma Wedgewood. They have ten children: six boys, four girls.

1842 Moves to Down House in Kent.

1859 Publishes *On the Origin of Species*.

1868 Publishes *The Variation of Animals and Plants Under Domestication*.

1871 Publishes *The Descent of Man, and Selection in Relation to Sex*.

1872 Publishes *The Expression of the Emotions in Man and Animals*.

1881 Publishes *The Formation of Vegetable Mould Through the Actions of Worms*.

1882 Dies at home.

The Conservative Revolutionary

> If you look at history's geniuses, they tend to be a very prickly lot. I do not think Newton was a very nice man. Galileo certainly was a driver. I am not sure I would have liked most of history's geniuses – I would have admired them, I would have stood in awe before them. But Charles Darwin was the most kind and genial man. It was not just the veneer of British upper-class charm, he lived that way. He is a man of remarkably admirable qualities. He also lived a fascinating life.

IN THOSE easy but high terms, the American palaeontologist and popular science author, Professor Stephen Jay Gould, praises Charles Darwin, whose fascinating life produced one of the most revolutionary ideas in the history of science. Darwin's theory of natural selection radically challenged the nineteenth century's view of the universe and everything in it. Instead of an ordered world created by God, Darwin asked his contemporaries to accept blind competition and random variation, where species struggled to exist and only those best adapted through sheer good fortune survived.

It is such a radical although simple theory, and many people today still reject what they think are its awful consequences. Yet for biologists, doctors, cosmologists and philosophers, Darwin's work has, if anything, become even more important than it was a hundred years ago. Darwinism today is the area where many of the most interesting scientists seem to gather. Philosopher Daniel Dennett, Professor of Arts and Sciences at Tufts University, Massachusetts, can scarcely pitch too strongly for its vital contemporary relevance.

> If I could give an award for the single best idea anybody ever had I would give it to Darwin, because his idea just unifies in a stroke these two completely disparate worlds, until then, of the meaningless mechanical physical sciences, astronomy, physics and chemistry on the one side, and the world of meaning, culture, art and of course the world of biology. One stroke shows how to unify all the sciences.

Richard Dawkins, the Oxford Professor for the Public Understanding of Science, has done more than most this century to reinterpret and explain Darwin to a new generation. He, too, sees him as 'a tremendously nice man' but is in no doubt also that:

> his achievement was tremendous.
>
> He discovered a principle which with hindsight seems enormously simple, but it is hard to believe that anybody did not think of it before, yet nobody did, not really. So it had to wait until the middle of the nineteenth century, which seems awfully late compared to, for example, Newton. And on the face of it at least I think it seems a much more difficult achievement. He is, undoubtedly, my hero.

Charles Darwin was born to a wealthy upper-class English family in Shrewsbury, the fifth of six children. His mother, Susanna, was the daughter of the famous Wedgwood family, his father, Robert, was a wealthy doctor. The biographer Janet Browne, Reader in the History of Biology at the Wellcome Institute for the History of Science, believes the best insight we have into the young Charles' privileged background is not from science, nor even from his distinguished intellectual genealogy, but from literature.

> He was very much a child that one could imagine stepping out of the pages of Jane Austen – not so much *Pride and Prejudice* but almost *Emma* – the picnics, dances, assemblies. Darwin spent most of his teenage years riding, shooting, hunting and collecting. He collected stones, he collected beetles, he even says at one point he might have collected biscuits, because he was just interested in acquiring

> things, in the way that lots of teenagers are. One of the most fascinating things about Darwin is this quirk that he was not an obviously brilliant child. His friends all remembered him as being a perfectly nice boy, but not someone you would have singled out straightaway as being destined for greater things; not an intellectual.

There are several intriguing links between the scientists in this book. Several – in accordance with their culture but also to a quite inordinate degree – were spellbound by religion and often by religion of the most text-bound and fundamentalist kind. Others were as apparently absent-minded, unworldly and obsessive as Archimedes. Some, like Darwin, showed few if any early signs of the superlative later insights which would provoke others to call them geniuses.

Stephen Jay Gould, though not Darwin's biographer, has made himself an authority on the man as well as the theories. To put it at its simplest – and, I think, truest – he delights in Darwin.

> He basically spent his years at Cambridge gambling with his wealthy friends. His father was so annoyed at one point he said he would only be good at hunting and rat-catching and he would be a disgrace to his family, and then somehow he found himself on the ship, the *Beagle*. He always had a passion for natural history, that was the one thing that kept

> him going through the years of formally academic non-interest, so there was always the passion for science beneath it all.

Darwin himself would not have quarrelled with that judgment. He abandoned plans made for him first to become a doctor like his father, then to be a clergyman. Perhaps he would have continued in this rather aimless way were it not for an opportunity, which he seized aged twenty-two, to travel round the world on board the *Beagle*. It is significant, though, that the seemingly aimless young Darwin suddenly saw an opportunity he wanted quite violently. Yet his famous voyage almost failed to happen – what a world might have changed there. Captain Robert FitzRoy, who was looking for a naturalist to join him on his nautical survey expedition, was not at first keen on the young man who presented himself. FitzRoy was a great believer in physiognomy and he thought the shape of Darwin's nose made him a rather disreputable character. Darwin persuaded him otherwise.

Then he had to face his disapproving father. In despair, Darwin wrote down each of his father's objections to the voyage in a letter to his uncle, Josiah, who provided comments on each point so that he could confront his father's objections – all eight of them.

1. Disreputable to my character as a Clergyman hereafter
2. A wild scheme

3. That they must have offered the place to many others before me, the place of Naturalist
4. And from not being accepted there must be some serious objection to the vessel or expedition
5. That I should never settle down to a steady life hereafter
6. That my accommodations would be most uncomfortable
7. That you should consider it as again changing my profession
8. That it would be a useless undertaking

Darwin was not to be put off. Together he and Josiah changed his father's mind. In December 1831 Darwin excitedly set off on a voyage that was to take him around South America, to Tahiti, Australia, Mauritius, Africa and, of course, most famously, to the Galapagos Islands.

For much of the journey Darwin was seasick and confined to the cramped quarters he shared with the equally young FitzRoy. But he took advantage of every opportunity to visit the strange new lands where the ship docked, formulating theories about everything, from coral reefs to plankton to mountain ranges. He sent hundreds of specimens of animals, fossils and rocks in large cases to friends and colleagues back in Britain.

Darwin was untrained, except for his misspent youth collecting things, but his work was prodigious and his discoveries many. However Professor Richard Darwin Keynes,

Darwin's great-grandson and the editor of *The Beagle Diary*, notes that Darwin did not always get it right.

> Of course what he said in *On the Origin of Species* and what he wrote in his diary afterwards, was that the important things were the fossils that he found and then the birds on the Galapagos.
>
> But as a matter of fact what had happened was that he did not really take any interest in the finches when he was on the Galapagos. He did not actually record nearly as well as he normally did which particular island he collected them from. And the whole point was that there were a whole family of birds which were not all finches. Some of them looked like warblers and some of them looked like other sorts of birds. It was only when the specimens were given to the ornithologist John Gould at the Geological Society that Gould pointed out how interesting they *all* were. Then Darwin got in a great flap and had to go round trying to sort out his specimens and trying to get more specimens from some of the other people like FitzRoy and some of the other members of the crew who had also brought some.
>
> He was told by the acting governor of the Galapagos, who was actually an Englishman, that the Spaniards could tell from which island any one of the tortoises came from the particular shape of its shell, and Darwin duly reported that, but he did

> not actually collect any specimens of the tortoises while they were there. All he did to the tortoises was to eat some of them. They were quite good for making soup, he said.

Hindsight is a great gift, but the initial confusion facing Darwin in the matter of specimens is not entirely surprising. So rapid and feverish were his opportunities for collecting and so quickly was he making astonishing connections and discovering new insights that it would have been much more suspicious had everything been as neat as ninepence after that extraordinary voyage. Without much difficulty, he sorted out what he needed. His peers had no doubts of his achievement.

On his return Darwin, now a famous naturalist whose discoveries had been reported in the newspapers in Britain, decided to marry. Drawing up another of his lists, he rather unemotionally spelled out the pros and cons. Under the column 'Marry', he wrote:

> *Children – (if it Please God) – Constant companion, (and friend in old age) who will feel interested in one – object to be beloved and played with. Better than a dog anyhow – Home and someone to take care of house – Charms of music and female chit-chat – These things good for one's health – But terrible loss of time.*
>
> *My God, it is intolerable to think of spending one's whole life like a neuter bee, working, working and nothing after all – No, no won't do – Imagine living all one's*

> *day solitary in a smoky dirty London house. Only picture to yourself a nice soft wife on a sofa with a good fire and books and music perhaps – compare this vision with the dingy reality of Grt Marlbro' St. Marry, marry, marry. QED.*

The list is hardly a contender for political correctness, but it prompted him to an unusually happy, stimulating and fertile marriage with his cousin, Emma Wedgwood.

For the next twenty years, Darwin – a man of independent means – sat and thought about the theories which had first come to him as he had sailed around the world. In the 1840s he moved to Down House in Kent, stayed there for the rest of his life, and never again went abroad. In the one short swoop he had gathered everything for a whole lifetime of intense thought. I asked Stephen Jay Gould why there was this long gap between his voyage, which was supposedly so influential, and the publication of *On the Origin of Species* in 1859. His explanation surprised me: in one sense it could be called the triumph of mind over mere matter.

> What he saw on the *Beagle*'s voyage very much prompted what was in his mind. I do not mean it was irrelevant but the development of the theory of natural selection was done in London, largely in the library of the Athenaeum. He read poetry, he read economics, he read science, he read philosophy, he read the entire canon of the Western thinking and he brought it all together into this theory of natural

> selection, not as a simple kind of obvious plotting induction from facts of nature that he had seen, but as a brilliant synthesis.

I was fascinated by this response. What, I wondered, had he got from all this varied reading? How, I asked Gould, did it add to his observations on the voyage of the *Beagle*?

> He wanted, for example, to understand the nature of science, so he read Kant and other philosophers and sociologists. It was a directed reading campaign. He in a sense knew what he was after and he knew it was a major reformulation. It was the species problem.
>
> He did not invent evolution. There were many evolutionary schemes. He thought that of Lamarck was fatuous.

Jean Baptiste Pierre Antoine de Monet Lamarck (1744–1829) was a French naturalist and a key scientist in early evolutionary thinking. Unlike Darwin, he believed that evolutionary change was caused by the environment and that such changes could be inherited. His most famous example is the giraffe's neck, which he believed to be long because giraffes had spent centuries reaching up to the taller branches on trees to find leaves to eat.

As Gould explains, Darwin was suspicious not only of Lamarck but of all previous evolutionists.

He did not think any of the previously proposed evolutionary schemes had ever had workable mechanics. What he ended up with was so philosophically radical. In most Western traditions, if you are going to believe in evolution, you are going to try and discover some complex force in nature which will automatically produce it as a result of its own nature, whereas with Darwin you had this bizarre mechanism that works through a hecatomb of deaths. If you want to get from A to B you kill off 99.9 per cent of the species that still has feature A and you differentially preserve that thin fraction that is varying accidentally in the direction of B. It is a bizarre mechanism and yet it does seem to work.

I asked Gould if he would explain and encapsulate Darwin's theory of natural selection.

The standard encapsulation is that, at least in its bare bones mechanics, the theory of natural selection is just three obvious facts of nature plus an almost syllogistic inference.

First, that the species and organisms produce far more offspring than could possibly survive. The cod, as Darwin points out, lays millions of eggs. If they all survived the oceans would be filled up with cod in six months and they would pile up on land at a very alarming rate. Darwin carried on the theme for the slowest breeding creature, the African elephant. If they all survived the continent would soon be shoulder to shoulder in elephants, so that is certainly true.

Second, that all organisms vary. That is just folk-wisdom – if you look around any room we are all different.

Third, that the sum of that variation is inherited, because natural selection is a genealogical theory. It is not going to do you any good to have differential reproductive success if your offspring are not more like you because you are trying to pass more of what is distinctively you into the future.

Put those three together – that only some can live because all organisms over-produce, and there is variation in the variations inherited – then you just make the inference of natural selection, and since only some can survive, on average those that survive will be the ones that are better adapted, better suited, better fitted to change in local environment.

Let us cite a ridiculous caricature, but it works: the elephant. Say there is a population of ordinary elephants in Siberia when it is warm and then an ice age starts. Well, there is natural variation in the amount of hair of these elephants on average, and the hairiest elephant can still fall in a crevasse, but on average the hairier elephants, because it is getting cold, will do better and have more offspring and a hundred generations down the line you get a woolly mammoth.

It may seem simple to explain but the impact of *On the Origin of Species* shook society. On Saturday June 30th 1860, barely six months after publication, a famous and acrimonious debate on evolution took place at the meeting of the British Association for the Advancement of Science in Oxford. The Bishop of Oxford, Samuel Wilberforce, demanded to know whether he was descended from an ape on his grandmother's or his grandfather's side.

Darwin himself was not at the meeting but his friend, the botanist Joseph Hooker, immediately wrote to Darwin to tell him what had happened. Darwin replied:

July, 2nd 1860

My dear Hooker,
I have just received your letter. I have been very poorly, with an almost continuous bad headache for forty-eight hours, and I was low enough, and thinking what a useless burthen I was to myself and all others when your letter came, and it has so cheered me; your kindness and affection brought tears to my eyes . . . How I should have liked to have wandered about Oxford with you, if I had been well enough; and how still more I should have liked to have heard you triumphing over the Bishop. I am astonished at your success and audacity. It is something unintelligible to me how anyone can argue in public like

orators do. I had no idea you had this power. I have read lately so many hostile views that I was beginning to think that perhaps I was wholly in the wrong, and that the Bishop was right when he said the whole subject would be forgotten in ten years: but now that I hear that you and Huxley will fight publicly (which I am sure I never could do) I fully believe that our cause will in the long run prevail.

The philosopher Daniel Dennett believes we cannot overestimate the impact of natural selection on Darwin's contemporaries.

> When I try to imagine how puzzling, how counter-intuitive, how strange and threatening this idea seemed to some people, I conjure up a little fantasy. Suppose some scientists came out tomorrow and said 'Well, here's what we've discovered: there's been evolution among numbers. You know number seven wasn't always a prime number, it started out as number four'. You think, 'This is crazy!'. As far as I know, it is crazy, but that reaction of frank, indignant disbelief was part of the negative reaction to Darwin's ideas. So he had to overcome that and he also had to overcome the fear that people had that he was demoting all that they held dear.

This outcry despite the fact that Darwin himself had hedged

his bets. He did not coin the term 'evolution' himself and he never endorsed the idea of progress which his writings seemed to suggest. Always a cautious man, he actually said very little about the subject which so upset his opponents. Stephen Jay Gould takes the story on.

> Darwin was clearly a philosophical radical and he revelled in it. He loved tweaking the nose of Establishment thought on the issue of inherent progress. On the other hand, politically he was Liberal and socially he was quite conservative – I do not mean in belief, I mean in lifestyle. He loved being a country squire, he loved genteel manners. He did not like the lifestyle of radicalism although he liked some of the ideas. He did not like them in his house though he liked to read their books. All these are interesting conflicts, and because of that he was very sensitive to generating too much upset with ideas whose upsetting character he revelled in.
>
> There is the paradox. So he does not publish. He developed the theory of natural selection in 1838 and did not publish it until 1859. I think he was afraid to expose – not evolution, that was a common unorthodoxy – he was afraid to expose the radical implications of his own view of it, and then when he published *On the Origin of Species*, he did not say anything about human evolution; he has one little throwaway line and a few other references here and there. He says: light will be thrown on the

> origin of man and on his history – that is all he says about human evolution.
>
> He got a little more courageous and he modified it in subsequent editions of *On the Origin of Species* and he wrote *Much light will be thrown on the origin of man* and then in the early 1870s, when he finally published a rather speculative book *The Descent of Man* on human evolution, he did get round to it.

During this long delay, or long gestation, into which so much has been read from so many different perspectives – including the notion that his many ailments were caused by the psychological effect of withholding or repressing his revolutionary discoveries – Darwin was sitting on the fence, according to Janet Browne.

> His wife was a religious woman. He too still possessed some residual sense, in the way that English gentlemen of his period did, that there was something greater than himself, that there was a deity in some way, and many of his closest friends were also religious people.
>
> So he did not want his book to slap them in the face and yet his theory necessarily requires that you do not include any kind of Godly design or initial creation or any divine supervision at any point. His theory was utterly naturalistic. So he sits on the fence all the way through that book. He was a very clever writer.

If Darwin sat on the fence and declined to debate the question of human evolution or whether there is a God, his successors have been less reticent. The difficulty for many people has been the question: if human beings evolved then are we still evolving and if so how can we tell? Richard Dawkins offers an answer.

> If you look back two million years in human history, the big thing you notice is that the brain is much smaller and so if we are to simply extrapolate evolution two million years hence, you would expect that we would all look like the Mekon – great big bulging brain. I do not think that will happen. For one thing, you can never expect evolution to go on extrapolating in the direction it has already come. Humans are a particularly difficult case because we are no longer subject to the sort of ruthless cutting edge of natural selection that our ancestors were. It is hard to die young now. So to the extent that there is natural selection going on, it is not survival that is at stake, it is just reproduction. That is important and Darwin recognised that in his other great theory – the theory of sexual selection. There are all sorts of reasons why some individuals might reproduce more than others, and if any of those reasons have a genetic component then, by definition, that is natural selection, but it would have to be sustained for a long time, hundreds of thousands of years

> perhaps, to have a major effect. So you have to say, for example, if you wanted to see the brain growing, the brainiest individuals would have to have more children than the less brainy individuals. There is no sign of that and there is no particular reason why they should. My suspicion is that even if there is a genetic component now, it is unlikely to be the same genetic component say even in a few centuries' time, let alone hundreds of thousands of years.

On another point – one which again comes howling out of the Darwinian debates – I asked Richard Dawkins whether, if we razed the planet clean and started again, *Homo sapiens* would inevitably turn up or would it be purely a matter of luck.

> *Homo sapiens* would not turn up, that is just too implausible. If you rephrase the question, would a brainy animal turn up? Would language turn up? Would a bipedal brainy animal turn up? Then it is less uncertain. However I would say that although bipedalism has turned up – dinosaurs, birds, kangaroos – extreme braininess appears really only to have arisen once; language has certainly only arisen once, unlike, say, the eye which has arisen forty or fifty times independently in the animal kingdom. It looks as though eyes evolve at the drop of a hat, contrary to many of Darwin's critics who

> thought they were very difficult things to evolve. But brainy things and, especially, language, in the face of all this time, only evolved once. So I should have thought it is highly unlikely that it would evolve again.

Although the idea of our being at the top of an evolutionary pyramid was displaced long ago, nevertheless our interest is excited by anything which explores or explains our own particularity. I asked Richard Dawkins how he accounted for what I thought was the slow but unique arrival of 'braininess'.

> It is not that slow, it's pretty fast as fossil histories go. There are several theories. Perhaps braininess was sexually attractive, or the ability to remember sagas, dance steps, or all sorts of other possibilities. Another theory is to do with social scheming – outwitting and outdoing is very important: tribal politics, braininess at a premium there.
>
> My own preferred theory is something that I call software/hardware co-evolution, derived from a computer analogy. The computer has evolved at the same high rate as the brain has biologically. Cultural evolution is much faster, and I think there is some sort of arms race, co-evolution meaning an upward spiral of software advances, which means the hardware has to keep up. This opens the way to further software advances and so it escalates. It

happened in the computer industry, and that might have happened to *Homo sapiens*.

Think of language arriving two million years ago, a software advance which immediately provided advantages for individuals good at it – there was intense pressure for brain size to increase. Then once the brain size had increased that opened the way for language to improve. It may not have been language, that may have come too late, but I would put a small amount of money on some software advance being the trigger.

Perhaps this would seem to some a rather bleak view of how human beings evolved on earth and maybe it explains why so many people still do not accept Darwin's theory a hundred and fifty years after he first formulated it. I turned to the renowned evolutionary thinker John Maynard Smith, Emeritus Professor of Biology at the University of Sussex, to investigate the strength of this resistance.

It is my impression that many people are very reluctant to accept the implications of natural selection. They would much rather believe that they were not the product of rather blind mutation and rather brutal selection and had been specially made for the job.

I asked him if he thought that the idea of the universe

progressing towards and leading up to us was dying very hard indeed.

> I think it is dying extremely hard. It is not within biology, it is as accepted as say the atomic theory is in chemistry, but outside biology and even with other scientific colleagues I am finding myself constantly having to explain. I think it has to do with the fact that Darwinism is an explanation of how we came to be here. It is a story about origins, if you like. Now every civilisation has some account of origins, like Genesis, but the function of these stories is usually to tell people that they are at the centre of the universe, that God loves them for one special reason, they had a special day laid aside to make them and so on, and they do not like a theory which does not give them any special place in the universe, that says they are just one animal out of millions. They do not like it.

People are particularly resistant, myself included, to the idea that human creativity – that is, poetry, language or art – can all be explained by Darwin's theory of natural selection. Daniel Dennett told me he has no such doubts.

> I often hear humanists, literary people, poets, say 'Well, it's all very well to have a Darwinian explanation of the design in plants and animals, but certainly not a Darwinian explanation of the

> creativity of say a great poet', and I say 'Wait a minute, you're telling me that you have no trouble with the idea that a blind, mechanical, thoughtless process can account for a nightingale but cannot account for an *Ode to a Nightingale*. You think an *Ode to a Nightingale* is that much more wonderful as a designed thing than a nightingale itself. I just don't think you know enough about nightingales. Certainly if natural selection can explain that amazing bird, it ought to be able to explain an ode to that bird.'

Stephen Jay Gould differs from what could be called this strict Darwinian view of all things including culture.

> The terrible mistake that the hyper-strict Darwinians make, like Dawkins or Dennett or Maynard Smith, is to assume that, because something has its origin in the process of natural selection, and that human consciousness arose by natural selection – and I presume it did – therefore everything we do as a conscious agent has to be explained as a Darwinian adaptation. That is not true, that is silly.
>
> Let us say that natural selection is the process whereby the human brain got to its enormous size in nature – having developed that large brain for a set of reasons that we needed in our African savannah ancestry. Whatever they were, the point is that the brain is the largest computing device that

> evolution ever built. It is capable by virtue only of its structure of doing a whole range of things that have nothing to do with the adapted reasons for its original attainment of this large size. After all, the human brain did not get big so that we could read or write, or think about Darwin, or try to figure out natural selection theory. All of these are not adapted side consequences of developing this computing power for other reasons, therefore most of what we call human nature, the universals that we glory in or fear are in fact probably non-adapted side consequences of building a large brain for other reasons.

I asked John Maynard Smith, one of the fervent Darwinists, whose side he thought Darwin himself would have been on.

> Well, it is hard to say. Obviously both Gould and I would each like to have Darwin on our side – who would not? My feeling is that Darwin was essentially an adaptationist. Gould is only an adaptationist on Mondays, Wednesdays and Fridays and I am an adaptationist all the time.
>
> I like to think that Darwin on the whole would be on my side. On the other hand, let us face it, Darwin was essentially not a mathematician – as I am. The essential point about Darwin is that if you look at organisms, either at the biochemistry

> of their structure of their behaviour or anything else, the more you look, the more you are struck by the fact that these structures are *for* something. They look as if they were designed to do something and the only mechanics we know that will account for that is natural selection. So without natural selection, we would be just completely bewildered by biology I think.

I wondered, given that I had been told more than once what Darwin did not know, what John Maynard Smith would tell him were he alive today.

> I have often wondered that. I would want to tell him first and foremost about genetics, I think, because the one thing that he did not know about, did not understand, were the laws of inheritance and the method of transmitting information between generations through our genes and so on. He would love to know it, I think, and I believe one could explain it. That would be the main thing I would want to tell him about.

I came back to the question I had raised with Gould and Dawkins and asked Maynard Smith if he saw Darwin as being at all useful as a way of understanding cultural change, using culture in an artistic sense just for the sake of this particular question – in the arts, in literature, history, painting, and so on.

I am not sure. The notion that ideas are inherited – it is Richard Dawkins' notion of a meme, if you like – has something going for it and I think one can begin to think about culture as a kind of selective process, but I am not enormously sold on that idea. It is a possibility, it is worth trying. I do not have any ambition to see Darwinism take over the arts or history or cultural analysis or those things. Let me be personal for a moment. My younger son is an artist – a performing artist. He and I are seriously thinking of doing some research – I do not mean cutting things up or measuring things, but talking to one another and watching one another to compare the way he behaves and the way I behave – the way biologists try to explain the generation of form and shape and so on, and the way artists do. But I do not for one moment imagine that I or Darwinism is going to take over what he does so it can be interpreted. Maybe he will take over Darwinism and explain what we do.

Darwin died in 1882 at the age of seventy-one. Stephen Jay Gould believes he never lost his great insight and brilliant originality, but I wanted to know why Darwin chose to spend the last years of his life writing a book about earthworms.

It is so characteristic for a scientist, not only great geniuses but people who fancy themselves geniuses (which is a much larger class), to try to write a summative book late in their lives: highly speculative, philosophical, invariably fatuous and an embarrassment for later generations.

What Darwin did was so marvellous. He took the results of forty years of experiments he had been doing and some early papers on earthworms and he lovingly writes this book called *The Formation of Vegetable Mould Through the Action of Worms*, published in 1881, a year before he died. He shows how worms, whom we tend to disregard because they seem so insignificant and humble, are responsible for these enormous earthworks. So the earthworm is just an elaborate metaphor. This humble little creature putting solids through its body, millions of them over thousands of years can do an immense amount of work and so too can evolution. It is a brilliant summation and it was true to his principle that it is not fatuous ideas that will change the world, but it is patient and humble understanding of how nature works.

Jules Henri Poincaré

(1854–1912)

1854 Born in Nancy, France.

1873 Enters the Ecole Polytechnique.

1875 Studies to become a mining engineer.

Date unknown. Marries. Has three daughters and one son.

1879 Becomes Professor of Mathematical Analysis at Caen. Receives a doctorate from the Ecole Normale Supèrieure des Mines.

1881 Becomes Professor at the University of Paris until his death.

1887 Elected as member of the *Académie des Sciences.*

1889 Publishes paper on 'The Three-Body Problem', which wins the prize offered by King Oscar II of Sweden and Norway. Made a knight of the French Legion of Honour.

1903 Publishes *Science and Hypothesis.*

1905 Publishes 'Sur la dynamique de l'électron', which anticipates Einstein's theory of special relativity, and *The Value of Science.*

1906 Elected president of the *Académie des Sciences.*

1908 Publishes *Science and Method.*

1909 Becomes first scientific member of the *Académie Française.*

1912 Dies, following operation.

The Man Who Discovered Chaos By Accident

We are better at predicting events at the edge of the galaxy or inside the nucleus of an atom than whether it will rain on Aunty's garden party three Sundays from now, because the problem turns out to be different. When you pushed the numbers through the computer you can see it on the screen. The future is disorder. A door like this has cracked open five or six times since we got up on our hind legs. It is the best possible time to be alive, when almost everything you thought you knew is wrong.

EVERYBODY HAS heard of chaos theory. Tom Stoppard's popular West End play *Arcadia*, from which the above extract is taken, intrigued and entertained many non-scientists like myself. This weird scientific phenomenon has inspired the arts from computerised pictures to the peculiar sound of fractal music.

Yet most of us, I suspect, would imagine that it is a late-twentieth-century discovery, whereas the original theory was first formulated over a hundred years ago by Henri Poincaré, the great nineteenth-century French mathematician. He is the least known of all the scientists in this series, but is a figure of enormous importance, argues Ian Stewart, Professor of Mathematics at Warwick University.

> 'I think he is my favourite mathematician in many ways. You have to picture a figure with a goatee beard, heavy spectacles, not very well co-ordinated, very focused on his mathematics. Poincaré was unusual because he was also interested in science in general, in culture in general. He was very broad as well as deep. There are many stories like someone came to visit him and Poincaré forgets the person waiting for him and carries on working and eventually he rushes out and says 'Go away'. Now that may not be absent-minded, it may actually be very clever!
>
> He allegedly failed many exams when he was young, including a mathematics question he got wrong. He got zero for his drawing, which nearly meant that he completely failed the exams he was taking. Poincaré was completely hopeless at drawing, not because he could not visualise, but because of his hand–eye co-ordination. He was not co-ordinated enough to put down on paper what his mind's eye

> could see. So he is slightly eccentric, but very, very bright.
>
> The fact that he could not do simple mathematics is not uncommon. Most mathematicians are pretty awful at arithmetic, their minds are on higher things and they are so convinced that arithmetic is straightforward and easy that they are sloppy and they actually get it wrong. Poincaré several times explicitly says that when he does calculations or similar things he is likely to lose track halfway through and not notice it. So he does not have a great deal of confidence in the symbols on the paper. What guides him is whether the whole thing makes sense, and actually this is a good principle because it is so easy to make a silly mistake, but if the whole thing looks coherent, then it is much more easy to feel confident in it.

Henri Poincaré was born in 1854 into a wealthy and distinguished family. His first cousin, Raymond Poincaré, became Prime Minister and President of France during the First World War. The young Poincaré was a sickly but brilliant child, who was taught at home until he was eight years old. Dr June Barrow Green, Lecturer in Mathematics at the Open University, takes up the theme of his absent-mindedness.

> One of the characteristics which people often

> associate with mathematicians, that of absent-mindedness, was apparent in him from an early age. He apparently used to forget whether he had had meals or not. There is a lovely story about him going for a walk with his mother and sister along a road where there was a stream. They crossed over one small bridge and he was left on the other side and did not notice and then when he saw them on the other side, he just walked in a straight line across to them. The fact that there was no bridge there had not occurred to him. So of course he landed waist deep in the water.

Ian Stewart considers why this sort of behaviour is common to mathematicians.

> I have a friend who says 'I am not absent-minded, I am present-minded somewhere else'. It would be nice to think of all these mathematicians being actually entirely urbane, normal types who were deliberately putting on this kind of eccentricity. I think it is because if you are a really top quality mathematician you spend an awful lot of time living inside your own head in a world that genuinely does not exist and your habits of thought get changed by that. And you probably do lose a little bit of contact with the outside world and with normal humanity and it just does not occur to you that you are doing anything strange.

In an introduction to Poincaré's popular book *Science and Method*, Bertrand Russell said that he thought that Poincaré was the greatest scientific mind then living, and there were some very great minds indeed at the time Russell was writing, including his own. Would his greatness be described as genius? Ian Stewart takes on the question.

> He is amazingly original. Poincaré single-handedly – I am exaggerating a bit – got at least two, possibly more, major branches of present day mathematics going from almost nothing, and he did not just get them off the ground. He created an enormous amount of the structure of these things, he really gave them a huge boost.
>
> One of them was topology. This is something which seems very remote from everyday concerns. It is a kind of very flexible geometry. It is a way of thinking about the really deep geometry of not just objects, but all sorts of mathematical structures. Without topology nowadays, mathematics could not function. Poincaré did not just come up with the idea that this was interesting, he invented a lot of the basic tools of the subject, and also some of the problems, including some which are still open today.
>
> Another one he invented is called 'dynamical systems'. In his day it was called 'qualitative theory of differential equations'. This was a really radical idea. This was the idea that you do not solve

> equations by coming up with formulae or lots of numbers, but you describe the general qualitative nature of the solutions. Do they get bigger and bigger and bigger until they become infinite? Do they stay within some fixed size? Do they go round and round forever, that kind of thing. And Poincaré found a way to describe the nature of the solutions without having any formulae to represent the solutions. Topology has created a very interesting way of thinking about the whole of mathematics in visual terms, but with a kind of analytic and algebraic back-up to make sure that you really are doing it right. The one that particularly interests me is his work in dynamical systems, in qualitative theory of differential equations. One of the best known things it led to, and this is very much Poincaré's creation, is chaos theory.

What is so intriguing about Poincaré's story is that he only discovered chaos theory accidentally. In the 1880s he entered a competition to try to solve one of the biggest mathematical problems of the day, the so-called 'Three-Body Problem'. It sounds simple enough: how do you determine the motion of three bodies in space if you know their starting points?

But like many mathematical problems, what looks simple is not. Finding the solution had foxed the greatest scientific minds of every age, including that of Sir Isaac Newton. Poincaré's solution was so long and so complicated that the other mathematicians judging the prize had to ask Poincaré to explain it to them. Poincaré wrote back with a commentary which ran to another hundred pages or so. Ian Stewart is fascinated by this.

> Poincaré was attracted into this particular area of chaos theory by a prize that was offered by King Oscar II of Norway and Sweden to celebrate Oscar's birthday. It was a mathematical prize, a money prize, less than half of Poincaré's annual salary, but the prestige was great. All the top mathematicians competed for it.
>
> The idea that Poincaré set out with was to prove the solar system is stable. This is an incredibly difficult problem, very subtle mathematically. Poincaré worked on this for three years, and it is clear from mathematical writing that at one point he really thought that he had proved it The memoir won the prize. Everyone got very excited.
>
> A bit later on the memoir was published in *Acta Mathematica*, one of the leading journals. It then transpires that the published version is different from the one Poincaré submitted. It differs in that the version he submitted was wrong. It is wrong because, completely uncharacteristically, he makes a geometrical mistake. He forgets a particular geometrical possibility – he is analysing cases that could occur in this problem and he misses one. Luckily he realises that this missing one is actually

much more interesting than he thought and it leads to a very interesting complicated kind of motion: what he realises, in effect, is that one of the bodies moves to all intents and purposes in a way that looks random.

It is not really random because you know the mathematical equations for it and if you ran it twice it would do the same thing again. But it *looks* as if it has got no structure. Poincaré at this point stops and says 'I do not know what to do with this'. This is in a sense horrifying, but it is there and he runs out of steam on the problem and does not know what to do with it. This is the phenomenon which later turns out to be called chaos. Chaos is when any deterministic system – that is a system that has no equations – has a solution that is so complex and so irregular that it appears to be random unless you know a lot of hidden information about what it is doing.

Other mathematicians picked up on this over the intervening sixty or seventy years and began to understand where this apparent randomness comes from, what this geometric case was like. It all turns out to be rather fascinating and wonderful. The word 'chaos' is really a bad name for the whole area, because it makes it seem as if it is horrible. In fact it is lovely, it is absolutely wonderful. It is full of all sorts of intriguing and forms and behaviours and things of that kind. Poincaré definitely created it. This was the first good example of chaos, the first one recognised to have this strange property. People developed it slowly but nobody, including Poincaré, recognised that this was the tip of a huge iceberg of very, very interesting mathematical questions and answers.

I put it to Ian Stewart that, as I understood it, Poincaré did not solve the problem and yet he is famous for what he did towards solving it. Ian Stewart was practically euphoric about it.

It is such a wonderful problem that failures to solve it, if they make sufficiently interesting strides forward, are perhaps more valuable than the answer itself. It is a representative of a whole class of problems saying 'How do you handle this situation?' The report saying why they were awarding him the prize says, 'He has not actually solved the problem but he has made such an amazing stride forward. He has created a whole new subject here, a whole new way of thinking.'

Sir Robert May, Chief Scientific Adviser to the Government and Head of the Office of Science and Technology, believes that Poincaré's work completely alters the way we see the world.

To say that it is a transforming insight is to my

> mind possibly going a bit over the top. But it is as transforming an insight as the original Newtonian dream that said the world is not driven by ghosts or goblins, it obeys rules. If the rules are simple then you get predictable outcomes and the things that appear to us to be random, the spinning of a ball in a roulette wheel, are like that because although there are rules they are just so many and so complicated. Chaos overturns that simple view of the world. It is ultimately no less than the end of that Newtonian dream, because what it says is, some of the simplest rules you can imagine, simple, rigid, prescribed rules, nothing random in them, can give you behaviour that is as random and complicated as anything you can imagine. Now that is wildly counter-intuitive and it undercuts the notion that complicated things are the result of hugely complicated rules.

Robert May is famous for his pioneering work in the 1960s applying chaos theory to biological systems. Scientists like him have been using Poincaré's insight in chaos to help figure out the laws governing the weather, disease and epidemics.

> The origins of chaos as we know lie in the work of Poincaré, but it lay pretty much buried in a cupboard. The physicists never wanted to get it out until very recently. It only moved centre stage

coming from two basic threads, one started by Edward Lorenz in meteorology, and another even simpler thread coming from population biologists in the seventies. From that it very rapidly moved centre stage.

The actual equations that give this weird behaviour had been independently discovered by four or five different people and their real importance was not recognised until a group of biologists including myself and some other people stumbled across them in the context of seeing them as metaphors for the behaviour of biological populations. And when we in turn independently rediscovered these bizarre properties, their real world implications were forced upon our attention because we were looking at them in a context of predicting population dynamics.

What makes Poincaré's discovery so important and so different from the clockwork view of the universe is its sensitivity to minute changes or what is commonly known as the 'butterfly effect'. This makes it very difficult if not impossible to make predictions. Supposedly when a butterfly in Tokyo flaps its wings, the result may be a hurricane in Florida a month later. Robert May develops this point.

When I was young one thought that the problem of making reliable long-term predictions of local weather was just a problem of computer power.

You had the equations, you had the local geometry of the earth and you had the weather satellites to give you the initial conditions and you just needed more number crunching to look further and further ahead. Now we realise that the basic equations are such that, depending on the weather today, as it were, they are going to be impossible to predict more than somewhere between ten and thirty days ahead for local weather because they have chaotic behaviour and the sensitivity to initial conditions makes it impossible.

In the world, say, of biological population, the examples are a bit trickier and a bit more ambiguous. I can show you examples in a laboratory of a fruit fly or a little water flea populations where I can grow them up and you cannot predict what the magnitude will be twenty days ahead under certain circumstances. Under some temperatures and conditions of growth they will be nice and steady and wholly predictable, like classic Newtonian dynamics. Under other temperatures and/or other environmental regimes, they will behave chaotically and, although the underlying equations are all simple, the population will not be predictable. It will be going up and down all over the place. Examples in the real world are harder to document unambiguously because most of the examples in the real world simply have too much going on, they are too complicated to be

> nice examples of very simple systems behaving chaotically.
>
> I would say we are still in a situation where most of what we are taught in schools and universities does conform to the Newtonian vision – most of what we are taught is that the world is orderly if the rules are simple enough, it is predictable and where it is complicated and unpredictable, like a roulette table, it is because it is a mess. All that is undercut. What we now know is sometimes if the rules are simple enough then it is simple, otherwise we could not build a clock, simple Newtonian clockwork. But the equations for the Newtonian clock, the force pendulum can sometimes give you nice regular clocks, but other times they can give you total, unpredictable chaos.

Chaos is not just complicated patternless behaviour. It is far more subtle. It is apparently complicated, apparently patternless behaviour that actually has a simple explanation. Robert May is exasperated at the way that chaos has become misunderstood in the century since Poincaré first stumbled across it.

> It has very rapidly been transformed from what it really is, which is a revolutionary insight that simple rules do not have simple consequences, into a vulgarised version that simply says the world is complex. You see that in its most simple form

in the appalling chaotisist in Michael Crichton's novel *Jurassic Park*, where you get this nitwit of a character who keeps running around saying chaos theory means the park will go wrong because it is complicated. Chaos theory does not say the world is complicated because it is complicated, it has no insights for politics because politics is inherently complicated. What it does say is the simplest rules you can imagine, with nothing random in them, can behave in so complicated a way that they do not have a predictable outcome.

As this is such a transforming theory, I asked Ian Stewart to attempt to assess how important it is now. His view differs from that of Robert May.

I think there is a message here for politicians. It does not tell them how to do it right but it tells them not to think that more and more rules and more and more elaborate rule books to cover more and more contingencies with less and less flexibility is going to solve the problem. It may actually create the problem.

For example, take the problems of over-fishing in the North Sea, off Canada, all round the world – there is a *lot* of over-fishing. Now there are political problems in enforcing the rules, but the rules themselves are often wrong because the big, funny, random-looking fluctuations in fish populations are partly to do with the way the rules themselves are applied. And so there are ecologists nowadays who are trying to convince governments that they need to rethink from the ground up the whole way that they manage fisheries. Not to be so naïve as to think 'If we just get the rules right, then the whole thing will settle down to something nice and regular and easily maintained'. It is not like that.

I think chaos theory is philosophically and politically liberating. That is why some people think it is dangerous and it is why it is too easy to exaggerate this side of it.

Regardless of whether anyone at the time understood the implications of his discovery, winning the Three-Body Problem prize in 1890 brought Poincaré public acclaim. From the age of twenty-seven until his death he was a Professor of Mathematics at the University of Paris and, in *fin de siècle* France, his name became renowned.

This eccentric, absent-minded scientist was an early populariser of science. His simple explanations of mathematics, physics and astronomy were translated into many languages. Poincaré was also interested in how mathematicians solve problems. His personal description, in *Science and Method*, of the difficult birth of a complicated mathematical idea became very famous.

Every day I sat down at my table and spent an hour or two trying a great number of combinations and I arrived at no result. One night I took some black coffee, contrary to my custom, and was unable to sleep. A host of ideas kept surging in my head, I could almost feel them jostling one another until two of them coalesced, so to speak, to form a stable combination. When morning came, I had established the existence of one class of Fuchsian functions . . .

Ian Stewart recognises the accuracy of description in this way of thinking.

It is still one of the best descriptions and the one that seems to ring true to an awful lot of people. Poincaré's idea is that you have to give your brain a really good shake up, exhaust all the obvious possibilities, exhaust even the less obvious possibilities, get to the point where you really despair of ever solving the thing and then go off and do something else. And he reckons the subconscious mind is then churning away trying different possibilities, coming up with other ideas, testing and doing all this work without your noticing, and at some point it is going to tap you on the shoulder and say 'Hey! I think I've got an idea'. And this very often happens or is triggered in some way by some completely extraneous events, such as stepping off a bus. I think there is a lot of truth in that. What is really going on may not be quite as he describes it but that is how it feels.

There is an analogy with writing, from the experience of writers I have read and, indeed, from my own experience. You work away hard on a piece of fiction and, at a certain stage, things happen that you had not anticipated at all, such as an unplanned character. It is not mystical or magical, it happens so regularly that there must be something else going on – and that must be the subconscious. It just seems more dramatic when it involves the solution of these great problems appearing in mathematics, much more concrete dramatic evidence of the subconscious at work.

With his combination of hard work and flashes of inspiration, Poincaré contributed to almost every area of mathematics. Debatably, Poincaré almost developed the special theory of relativity at the same time as Einstein. However, it is Einstein who gets the credit in a patriotic debate about who got there first. In Poincaré's native country, Michel Paty, Director of Research at the National Centre for Scientific Research, explained.

> It has been a matter of discussion – what has been the contribution of Poincaré to the theory of relativity? In fact, for a long time, the tendency has been to say that Poincaré failed in this respect and it is only Einstein who discovered the theory of relativity. Another more recent view is to say that Poincaré in fact discovered the theory of relativity at the same time as Einstein. Some even pretend that he discovered it before Einstein. In fact both of these interpretations are not true.

> People say Poincaré was not revolutionary. It is true that Einstein was brave and audacious. He was like a young man: he had young thoughts which immediately made a leap. But this does not preclude the fact that Poincaré made a large contribution to works in this field.

Ian Stewart is less equivocal than Michel Paty.

> Poincaré wrote down the equations; he obviously had in his mind very similar ideas. They were in the air at the time anyway. Einstein took a step that nobody else took and I think this is why Einstein deserves the credit for creating the theory. He said 'This is not just an interesting mathematical gadget, this is not just a way of thinking about certain problems in mathematical physics, this is reality, this is how it works, things really do get shorter when they move faster'. And he really put his reputation on the line by saying this because it is no longer just a little intellectual game. This is trying to tell people that reality is totally different from what they expect. Einstein was courageous enough to risk his reputation.

While he may not have staked his reputation on relativity, Poincaré was by no means a conventional thinker. He abandoned the more orthodox methods of solving equations with numbers in favour of visualising the answer. Poincaré himself believed that it was beauty rather than a desire to solve problems that inspired scientists to devote themselves to their work. He wrote:

> *The scientist does not study nature because it is useful to do so. He studies it because he takes pleasure in it, and he takes pleasure in it because it is beautiful. If nature were not beautiful it would not be worth knowing, and life would not be worth living. I am not speaking of course of that beauty which strikes the senses, of the beauty of qualities and appearances. I am far from despising this, but it has nothing to do with science. What I mean is that more intimate beauty which comes from the harmonious order of its parts, and which a pure intelligence can grasp . . . Intellectual beauty, on the contrary, is self-sufficing and it is for it, more perhaps than the future good of humanity, that the scientist condemns himself to long and painful labours.*

I came back to Ian Stewart on this point. Poincaré talked of mathematics and solutions being 'beautiful', and I had heard Ian Stewart talk in the same terms. The word 'beauty', I suggested, would surprise many people in connection with mathematics – pain would come to mind mostly. But Ian Stewart insisted on the word 'beautiful'. I asked him if he could explain how he and Poincaré employed it.

There is certainly such a thing as ugly mathematics.

> I think one of the reasons a lot of us experience pain when trying to do mathematics is that we are seeing the ugliness and are not getting the feeling of the underlying beauty. But there is a kind of deep beauty, and some of the really best mathematics in the world, when you understand it, gives you this wonderful feeling that it is elegant, that it draws upon all sorts of wonderful areas that you had not realised were connected. It is like a musical composition where, somehow, everything comes together and you get this feeling of complete inevitability.

Robert May agrees.

> Whether you are a pure mathematician, in it just for the sake of the game, or an applied theoretical physicist or a theoretical biologist using mathematics as a tool to understand how the world works, all such people will have a fairly shared and hard-to-define sense of what is really elegant and beautiful and what is just a mess. You would get it crystallised in questions such as whether Fermat's Last Theorem was really solved when the proof of it, after all, takes a book. Or whether you have really solved any problem if you have got to look at five thousand and eighty-six exceptional cases to prove it. That is not beautiful – it may get you the answer, but it sure as hell is not beautiful. There

> is a notion of beauty and it has been a guide in some areas of theoretical physics. It can be a guide, it can be a danger. To be captured by beauty as the criteria when you are looking at the real world can be very misleading in many biological situations where the real world is a mess.

Poincaré's own comments on the practice of science sound very modern. His essays are wonderfully accessible even today. What to me is so remarkable about Poincaré is the range of what he could cover, a different subject for every day of the week. Nor were his interests confined to mathematics. His first love apparently was natural history and he was the first scientist awarded membership of the *Académie Française* – a rare accolade. Today it is hard to imagine a scientist with such a deep knowledge of so many subjects. Ian Stewart concurs.

> He could work in pure mathematics, in number theory; he could work in analysis; he could work in applied mathematics with the many-bodied problem; he could switch from one to the other; he could see connections between all these things. In addition to this he could write popular science books that sold very well and said amazingly sensible things, often ahead of his time. There is nobody after him at whom you can really point and say they understood that much of the subject that deeply. There are two reasons for this. First, he is in many ways unique.

> Second, the subject just grew and grew and grew partly because of the innovations that he introduced.

I asked Ian Stewart if he thought that that part of Poincaré's genius came from having such a broad view and knowing about so many different fields of mathematics. Was that broadness of view and reach of mind important for the progress, for the development of mathematics? Without it, are people not going to go forward in the way that Poincaré took it forward?

> I believe that is true. I think in a sense the enemy of progress in not just mathematics is too much specialisation by too many of the people involved. I think most areas of science and mathematics are so broad that everybody has to specialise to some extent, but if all the specialists do say 'My speciality is the only thing in the world that matters and I am only going to work in that area', they are missing all sorts of potential cross-connections which are actually very, very fruitful. I am optimistic, I think mathematics is beginning to learn, but too much specialisation is a bad thing.

Robert May came at the question from a different angle.

> In Poincaré's time, the ambit of what was embraced by science was hugely smaller. People knew, to a good first approximation, nothing about the way

> living things put themselves together and interacted with other living things and so on. Today from the molecular machinery in the book of life, through to the working of that out in interactions among populations to the structure and function of ecosystems, we know so much more that simply to encompass the whole breadth of science is extraordinarily difficult.
>
> But by the same token, there are so many interesting opportunities to bring an insight from one area into another area that I would say we are much richer in people hopping across boundaries or creating new interdisciplinary subjects than ever before.

Henri Poincaré died suddenly at the age of fifty-eight. There were effusive obituaries in every newspaper, including one by Sir George Darwin, Charles Darwin's son, in the British press. They talked of the mathematician who had invented a whole new area of mathematics based on drawing, but whose own artistic skill was so poor that his classmates at school had to label his sketches because otherwise no one could work out what they were meant to be. Today, he is still honoured in France, where there are streets named after him. Michel Paty is not surprised that a relatively obscure mathematician is so honoured there.

> Since the French Revolution science and scientists have been considered as really important for

> the state and many scientists have had important positions, not only in the Academy, but in companies and as ministers. So I think that Poincaré is known by many people in the street.

Robert May and Ian Stewart differ as to whether the British will ever be as proud of their scientists. Robert May observes:

> In France there was recently a poll among schoolchildren and they asked who was the most famous Frenchman ever. Seventy per cent said Pasteur. I would be fascinated by the result of such a poll in Britain, but I bet you there would not be seventy per cent for any one person and it would not be a scientist.

Ian Stewart remarks:

> I was very amused recently to discover my name was used as a clue in the *New York Times* crossword, so in a sense you see we are getting into the culture. The French have had a very long respect for intellectual activity of all kinds. It is one of their great strengths. It comes out in funny ways. In France there is a very good market for popular science comic books, for instance. It is very hard to sell them in this country. I think intellectual activity is seen as a little bit secondary here, and

> I think that is a shame. But I am optimistic that there is still a lot of good will towards all kinds of intellectual activity here and that in the right circumstances it can all come bubbling up to the surface again.

By the end of his life Poincaré had received every honour, prize and medal available to a mathematician, and fifty-one nominations for the Nobel Prize in physics. The boy whose professor had called him 'a monster at mathematics' had proved to be just that, transforming the subject for the rest of the twentieth century. He may be relatively unknown to us but his contemporaries were in no doubt as to his genius. June Barrow Green has the final word.

> He had a terrific impact on the European milieu. The mathematicians in Paris, with whom he was working, recognised him as the greatest among them. Time and time again you read remarks made by those mathematicians to the effect that Poincaré was the best, and if you read any of the obituaries of Poincaré written by a lot of these very highly regarded mathematicians, it is absolutely clear that for them he was the genius amongst them.

Sigmund Freud

(1856–1939)

1856 Born in Freiberg, Moravia.

1860 Family moves to Vienna.

1873 Enters the University of Vienna to study medicine.

1881 Qualifies as doctor.

1882 Begins work at Vienna General Hospital, specialising in neurology.

1885 Studies with Jean Martin Charcot at the Salpêtrière Hospital in Paris.

1886 Returns to Vienna and sets up a practice as a doctor of nervous diseases.

1895 Publishes *Studies in Hysteria* (jointly with colleague Josef Breuer).
Marries Martha Bernays. They have three sons and three daughters.

1896 First use of term 'psychoanalysis'.

1900 Publishes *The Interpretation of Dreams*.

1904 Publishes *The Psychopathology of Everyday Life*.

1905 Publishes *Three Essays on the Theory of Sexuality* and *Jokes and Their Relation to the Unconscious*.

1908 First International Congress of the Vienna Psychoanalytic Society.

1913 Publishes *Totem and Taboo*.

1920 Publishes *Beyond the Pleasure Principle*.

1923 Publishes *The Ego and The Id*.

1930 Publishes *Civilisation and its Discontents*.

1938 Leaves Vienna for London when Austria is annexed by Germany.

1939 Dies in London of cancer of the palate and throat.

Science or Art?

SIGMUND FREUD'S study in the Freud Museum in Hampstead, North London has deliberately been left just as it was when he died in 1939. It gives a real sense of the man, a cultured man, a man of the arts and of science, but also a man who was the subject of much criticism during his lifetime and who has remained a controversial figure long after his death. It was there I met and talked to the psychotherapist and writer, Adam Phillips.

Perhaps it is easy to understand why Freud's work was shocking to people at the turn of the century, but why, I asked him, has his work remained controversial for so long.

> I think the controversies have changed over time. I think that, to begin with, there was the question of whether this could possibly be scientific and also there was the scandal of sexuality, the scandal, I would say, of children being sexual. I think by now, a hundred years later, the scandal is different, and it may not even be as glamorous as a scandal. We now have more of a sense of the context in which

> psychoanalysis emerged and therefore a kind of doubt about how universal these truths are. And also we have more of a sophisticated sense, simply having more accounts of analysis and more who have been analysed, of what kind of a practice this is. I think infantile sexuality has gone on being controversial and, of course, child abuse cases have, in a sense, brought this back to life.

Freud himself, in a rare recording made at the end of his life, declared:

> I started my profession as a neurologist trying to bring relief to my neurotic patients. I have discovered some important new facts about the unconscious and psychic life. I had to pay heavily for this bit of good luck.

Freud's revolutionary theories about the unconscious have, perhaps more than the work of any other giant in this series, influenced every aspect of our lives in the twentieth century. Yet rather like Darwin, there was little to suggest in Freud's background that this would be the case. Born into a comfortable Jewish family in 1856 in Moravia, then part of the Austro-Hungarian Empire, the Freud family moved to Vienna when Sigmund was four years old and he remained there almost continually until the last year of a long life. Adam Phillips believes Freud's Jewish background had a profound influence on him.

> It sounds as though he had what was then a traditional, middle-class, Jewish upbringing. That is to say he did learn Hebrew and his grandparents were rabbis. But by the time of the parental generation, I think, there was a lot of fear about the taint of being Eastern European. The implications of being Eastern European, I think, were to do with being vulgar and uneducated, uneducated by the standards of high Western culture. And I think that all the drive in Freud's parents was for Freud to become a respectable doctor.

Freud proved to be an able student and, after studying medicine, he chose to specialise in neurology, the study of disorders of the nervous system. An encounter with the world-famous Parisian neurologist, Jean Martin Charcot, who was studying hysteria, proved to be a turning point in Freud's career, as Adam Phillips stresses.

> Freud went to see Charcot in 1885 when he was twenty-nine, and he says in a letter that he went to learn new things and it sounds as though there was one of those perfect overlaps between somebody who was hungry for something, who is young, and somebody there ready to be found. Charcot was a star – that was the first thing. He had prestige and glamour. He was giving these demonstrations which were the theatre of Paris to some extent. And Freud was clearly looking for a way of moving

over from neurology and neuroanatomy into what we would now call fantasy and psychopathology. What Charcot was showing was that if you could hypnotise people into *having* hysterical attacks, you could by the same token take them away through hypnotism. So in a way he was showing Freud the power of words and this was all very much to do with fantasy. There were things of which people were unaware that they were enacting, that they were performing. So in a way he was showing the theatricality of what Freud would call neurosis. I also think that there was the element of sex. Charcot said it is always the genital thing. He did not market this, and make it part of his approach to hysteria, but there was a kind of black market of ideas around that really hysteria was about sexuality.

The other thing that I think was very important for Freud and Charcot was this idea of conversion. That hysteria was a conversion symptom, meaning that there was a sense in which people could not help but talk, and even if they could not speak, they would simply talk with their bodies. What I think Freud was getting a sense of, from his own work and from Charcot's demonstrations, was that there was an instinct to communicate, but that there were clearly catastrophes attached to speaking. So I think Freud was not saying that all we need to do is talk about sex. I do not think it was a revelation to anybody in the nineteenth century that sex was very important. What Freud was saying was important was the way in which sexuality resisted articulation and therefore got converted into what we might call body language or symptoms.

Freud returned to Vienna from Paris in 1886, a young man of thirty, and decided to set up a practice as a doctor of nervous diseases with a particular interest in the workings of the mind. Writing to his fiancée, Martha Bernays, shortly before their marriage, Freud is optimistic about his prospects.

I won't be able to write to you any more during my consulting hours because there is too much going on. The waiting room is full of people and I shall hardly be finished by three o'clock. The takings are not yet very brilliant, but the patients who avail themselves of my services are quite numerous.

Oliver Sacks, Clinical Professor of Neurology at the Albert Einstein College of Medicine, has made his own career out of studying and writing about cases rather like those of Freud. His books are widely read and highly acclaimed by his peers. I wanted to know why he thought patients were so keen to join Freud's new practice.

People did not come to Freud to be investigated, they came because they were tormented, because they were obsessed, because they were driven,

> because they were jealous, because they were frustrated, because they were depressed, because they were anxious, in some cases because they had strange symptoms which could not be explained by organic neurological disease. People came to Freud as patients to be helped. And psychoanalysis developed partly as a way of exploring what was going on with the patients and at the same time as a way of helping them.

Freud began to develop Charcot's new ideas about hysteria and became convinced it had psychological rather than physical origins. Gradually, he began to formulate analytical theories like the Oedipus complex, drawing on folklore and mythology. Susan Greenfield, Professor of Pharmacology at Oxford University, is a neurologist with a special interest in Freud and she explains why Freud felt that traditional science, the study of the brain, could not provide the answers to the problems his patients were bringing to him.

> Initially, he thought hysteria was due to a specific cause, a specific idea, but he gradually realised that, under hypnosis, some of the things his patients were telling him were actual fantasies, they were not real facts. He, himself, when he underwent analysis, realised that this Oedipus complex that he had identified was in fact identifiable in himself, even though it had no immediate cause – there was no history in his childhood of his mother seducing

> him – but nonetheless he had it. And that made him realise that these seeming-fantasies were part and parcel of the human mind and therefore one did not just have an abnormal cause, one simple cause like something terrible happening to you, a very clear-cut thing happening to you that caused a neurosis, but it was rather more complex than that.

I asked Oliver Sacks why Freud was so important in the history of neurology.

> He started as a neurologist and some of his early neurological books are still read a century later – for instance a book on aphasia in 1891. Also agnosia was devised by Freud at a time when analysis of conditions like this was very diagrammatic, just described in terms of centres of the brain.
>
> Freud had a dynamic view of all the different brain actions that were needed to form a sentence and to form language and which might break down. But probably it is the psychological theories and insights which remain especially pertinent to neurology. I think neurologists have to have a clear idea of the structure of dreams, of fantasies, of repressions, of the unconscious, of fixations, of complexes, because these things occur in their patients as well and clearly these things probably have a basis in neurology. Freud himself attempted to give us a sort of neurology of the mind in 1896

> but then he gave that up and realised it was far too early. But now a century later I think that Freud's idea for a neurology of the mind is beginning to become possible.

I was interested to learn what Oliver Sacks, as a neurologist working today, would regard as Freud's most important discoveries.

> Primarily his discovery of the unconscious, of the dynamic unconscious, of a whole area of feeling and thought which is inaccessible to consciousness and which is vehemently kept out of consciousness by processes of defence and repression. Freud was very much aware of what is sometimes now called the cognitive unconscious, for example, the way in which the mind deals with language unconsciously and comes out with a perfect sentence or perhaps comes out with a slip of the tongue.
>
> Freud had a dynamic view of the memory which was unusual at the time. In the 1890s, phonographs had just been invented, photography was very popular and memory was usually compared to a trace, some physical trace which would remain unaltered and inert until it was played. One of the things that Freud insisted upon was that memories get altered according to one's wishes and fears and further experiences and that they are in flux. There's now good biological evidence for the

> dynamic quality of memory. We are talking about someone who saw the complexity of the brain's organisation and its dynamism in a way none of his contemporaries did and also none of his psychiatric contemporaries did.

Yet Freud abandoned neurology, which has always been regarded as a crucial breakpoint in his career. Was this the turning away from science to something more akin to an intuitive system of healing? Oliver Sacks does not see it in those terms.

> I think he abandoned neurology because there was no way in the 1890s in which Freud could talk about the neurology of fantasy, of memory, of consciousness and there was no point in doing so. So I think he decided he would deal with the phenomena and that he would speak in purely psychological terms. I do not quite know how to put it, but I think he obviously felt that the reflex hammer and the ophthalmoscope and the tools of a neurologist were not going to allow him to study people's minds.

In 1896, Freud coined the term 'psychoanalysis' for his system of treatment of mental disorders. His stated aim was to turn neurotic misery into common human unhappiness, but his theories and methods provoked much hostility in the salons and cafés of turn-of-the-century Vienna. The conversations Freud was having with his patients were intimate, outrageous, even shocking. He asked them to say whatever came into their heads and found that they spoke more freely if they lay down on a couch while he kept out of sight. Caricatured beyond measure throughout the century, the original couch there, in Freud's study, in the Freud Museum was, to me, a strange sight. Adam Phillips explained why it was important for his patients to lie back and try to forget that Freud was there.

> Freud discovered that if you let somebody lie down and not see the person they are talking to, they are freer to get into their own internal delirium and it then becomes more akin to a dream state. So, in other words, what he discovered by implication is that when you face somebody you respond to their demand, the demand of their presence, in quite a different way. And I think the aim of psychoanalysis, when it was working, was that you would forget, in so far as this was possible, that you were in the presence of somebody else and you would begin to talk to yourself freely.

Was it out of that free association that the greatest self-discoveries and interpreted discoveries came, I asked?

> Yes, that intelligibility was defensive. And in order, as it were, to discover disowned parts of oneself, one had to be prepared almost to take the narrative

> out of one's stories, one had to be able to stop being a good storyteller and see what came into one's mind.

Both Adam Phillips and I had spent some time soaking up the sombre, powerful atmosphere of Freud's room in Hampstead. It was extraordinary to be in it, replicated we are told from the one in which he studied people for about fifty years in Vienna, with beautifully arrayed books in bookshelves, some of which he brought to London from Vienna, and the cabinets full of objects, masses of them, from Egypt and other ancient civilisations. It was as if in these objects he were exploring and unearthing Western civilisation, its roots and origins, just as he was looking for the roots of pain in his patients. The room is a time capsule. Adam Phillips agreed.

> Yes. And it gives a very powerful message to any patient who went in there. Freud is manifest in his objects and his books and, even if it tells you nothing directly, it tells you a lot evocatively about what such a man might be like.

The neurologist Susan Greenfield believes that Freud's ideas have permeated more areas of our lives today than most of us realise.

> Most people, if you said 'What were Freud's ideas?' would say 'Oh, it is all about the subconscious and

> how the subconscious influences your conscious states' or 'You know – you dream of a train going into a tunnel or phallic symbols' and I am sure there are lots of 'nudge-nudge' and 'wink-winks' if you show someone candles and so on. I am sure most people see the world in Freudian terms, although they might not admit it.

As his patients kept talking, more and more of their unconscious desires emerged. These desires seemed to be mostly sexual and sometimes destructive in nature but they were repressed as inappropriate by the conscious mind in everyday life. Susan Greenfield believes that Freud was right about our primal urges. But she puts a modern and less controversial interpretation on his theories.

> Freud's idea was of the power of the id, the most basic forces, atavistic forces that are there all the time. In order to find expression and legitimisation, they have to be catered for by an ego, a sort of consciousness.
>
> Now I would agree with Freud that there is within us a very primal, basic, sensational desire for a sensation which I would call pleasure and he would call sex: a positive feeling that is so primitive and so basic and so simple that it does not entail individuality; it is not the sort of pleasure you get when you have a pay rise or someone pays you a compliment, it is not like that, that is a kind of

> cognitive glow. This is the zap, the zap of sensual pleasure that comes from an intensive bombardment of the senses and I would like to think that that was what he was talking about when he was talking about childhood sex and childhood eroticism and so on. He was not talking about children rewriting the *Kama Sutra*. He was talking more about that pure pleasure feeling.

Oliver Sacks comes to this central question by a different route.

> Freud felt very strongly from the start that people were driven by powerful forces which were not entirely conscious. He used the term 'drive' and he saw these drives as being partly biological and similar to instincts. He regarded sexual drive, libido, as the most central and, in a way, its aberrations as the most productive of torment or difficulties.

I asked Sacks if this was one of the reasons why Freud's ideas were resisted so strongly – because of his insistence upon the centrality of sex and the sex drive and the excesses of the sex drive.

> Yes. It seemed outrageous, and when Freud threw in infantile sexuality as well that seemed even more outrageous, because one knows infants and children are innocent and not touched by sexuality. Of course, it was also resisted by many of those who

> worked with him, and Carl Jung and Alfred Adler and others first went along with this and then also rebelled at the notion of the primacy of sex.

To illustrate his theories of child sexuality, Freud used the Greek myth of Oedipus, who unwittingly killed his father and married his mother. Adam Phillips explains how Freud developed this ancient story.

> The Oedipus complex in a way formulates a very ordinary experience, which is that everybody has had the experience of feeling left out. Now Freud says this is a fundamental constitutive human experience, that you are born to a mother and father and to begin with it is as though you have a two-person relationship, there is something exclusive. And so you can live as though you are entitled to it all. This woman your mother, let us say, belongs to you, almost as though she is a part of your body.
>
> So you might begin to feel that you control her in some way, she is literally yours, and then it dawns on you that there is somebody else. This woman who really loves and adores you actually goes to bed every night with somebody else. Well, this is astonishing. You are suddenly confronted with something, which is that you are terribly important but that also there is a point of view from which you are not important at all.

Your mother is everything to you but you are not everything to your mother.

And I think this is some kind of very powerful shock that is recreated in states of sexual jealousy. So it is almost as though you go from being a couple to realising you are part of a triangle, and the question then is what do you do with this experience of feeling left out? You know that your parents have pleasures that exclude you and intimacies that exclude you. What do you make of that?

Well, one thing you can make of it is an attempt to spoil it. You can enviously attack it, or claim that you are not at all interested in those things, you have got higher things to do – like read books, say, whatever your particular pleasure might be. But whatever you do, you have to organise yourself around this fact, that these two people you love have a relationship that excludes you, and I think Freud realised in talking to patients but also from his own self-analysis that there was a sense in which, therefore, one was bisexual: one really did want to marry one's father and kill one's mother and the opposite. In other words there was a sense in which in order to desire somebody, it is as though you had to kill somebody else, you had to get rid of your rival. There was always a rival.

Such a theory remains controversial even today, but it

outraged society at the turn of the century. Throughout his life, despite his steadily growing reputation, Freud had to fight constantly to have his ideas accepted, not only among the public at large but even by the band of followers that was growing around him. He wrote:

> *From the year 1902 onwards, a number of young doctors gathered round me with the express intention of learning, practising and spreading the knowledge of psychoanalysis. The stimulus came from a colleague who had himself experienced the beneficial effects of analytic therapy. Regular meetings took place on certain evenings at my house, discussions were held according to certain rules and the participants endeavoured to find their bearings in this new and strange field of research and to interest others in it . . . Besides doctors, the circle included others – men of education who had recognised something important in psychoanalysis: writers, painters and so on.*

In 1908 the first international congress of the Vienna Psychoanalytic Society was held in Salzburg, and included great names of the day such as Carl Jung and Alfred Adler, though they did not join forces with Freud for very long. Alfred Adler impressed Freud in the beginning and became an early 'disciple', but the two never fully agreed on Adler's key idea, according to biographer Peter Gay, that 'Every neurotic seeks to compensate for some organic imperfection'. They finally split in 1911, Freud declaring that Adler was paranoid.

The break-up of Jung and Freud was much more acrimonious. Jung was Freud's chosen heir but he could not accept some of Freud's key tenets, in particular Freud's ideas on libido, including childhood sexuality and the Oedipus Complex. The letters Jung sent to Freud as the two began to fall out from 1912 onwards are, according to Peter Gay, patronising and angry. They accused one another of being unscientific, while Freud considered Jung was gullible and prone to believe in the occult. Jung resigned as President of the International Psychoanalytic Association in April 1920.

The jibe about being unscientific must have hurt Freud deeply. It was of the utmost importance for Freud that psychoanalysis be accepted as a science. He had been trained as a scientist and dreaded the accusation of mysticism, as Adam Phillips explains.

> In a sense, if you could not legitimate what you were doing as scientific by the standards of the time, you were then potentially in the world of the illegitimate, and I think there must be something here of Freud's Jewish history, because I think Freud's fear was that he would be seen to be like a medium or a psychic. In what sense is a psychoanalyst different from someone who gazes at crystal balls for example, or reads your palm? And the answer may be not very different, and I think there is a great fear about that.
>
> I do not think it is a science. It is not falsifiable. I do not think it is subject to any of the criteria of

> science. There are lots of kinds of truth and I am not saying that scientific truth is not valuable but it does not have to be the only kind of truth available. Lots of things that one loves and likes and matter most to oneself are not provable or falsifiable. You cannot evaluate a childhood, you cannot measure a childhood. It would be a simple-minded way of putting it, that an awful lot of the forms of, say, quantification simply cannot be done in psychoanalysis. You cannot measure, it is not subject to mathematical reduction, I think.

This is still a battleground – and of the utmost importance not least to those who have devoted their lives to psychoanalysis. Susan Greenfield has a different view from Adam Phillips.

> I think we have to put this in context. Freud was living in the last century. And what he may have regarded as science should be set against the then still popular idea of *natur-philosophie*, the idea of vitalism; that you had a life force that could not be reduced or explained.
>
> He was against them. His work was a retaliation against this glamorous but airy-fairy idea of life forces. So, seen against that, his work could have been called scientific and he was attempting, especially in the early days, to establish a cause for neurosis, a cause for hysteria. For him it was a

> scientific approach in that he was trying to establish a causality, not just throwing up his arms and saying 'Well, it is all *natur-philosophie*, it is all life force'. Now, by today's standards, of course, that does not measure up to the rigours of what we call science – objectivity, measurement, reproduction. So I think one has to be sensitive to the context and to the history of a hundred years ago, and what was called scientific then and now.

Oliver Sacks has no doubt that Freud's work deserves to be called scientific, especially if parallels are drawn with another revolutionary nineteenth-century thinker, Charles Darwin.

> Freud made a prodigious, patient, lifelong study of the way people think, feel, dream, act, and how they often do so in ways which are bewildering to them and actuated by mechanisms and ideas and feeling which were apparently not in consciousness. He moved from this huge mass of observation to certain general principles about the mind and in particular about its dynamic nature, and this is a scientific project. I think it is one which is analogous, say, to Darwin going to the Galapagos and going round the world, making a huge number of observations on the way animals live and evolve, pondering this for years and then writing his *Origin*. In a sense, Freud was concerned with an origin of

> mind, an origination of mind, rather like Darwin's *Origin of Species*.

Like Charles Darwin, Freud produced his scandalous theories from the confines of a comfortable marriage and a happy family. He was devoted to his six children, the youngest of whom, Anna, carried on her father's work after his death, becoming a world-famous child analyst in her own right. But his life was marred by family tragedy. The death of one of his beloved daughters and a grandson was followed shortly afterwards by the onset of agonising jaw cancer, which ultimately led to his death. The other shadow over the last years of his life was the rise of Fascism in neighbouring Germany. In 1933 the Nazis burned Freud's books in Berlin and by 1938, when they had annexed Austria, Freud finally resolved to flee with his family to England, where he spent the last year of his life quietly in North London. Frau Ingrid Scholz-Strasser, who manages the Freud Museum in Vienna, describes his departure from the city.

> In the entrance we have one of the huge suitcases which he used when he had to leave the country. Sigmund Freud did not want to leave Austria because he was already in his eighties and he was severely sick and he was deeply connected to Austria. Only when the Nazis arrested Anna Freud and took her to be interrogated, did he decide that he and his family would leave the country. His four sisters remained in Austria and could not be saved.

> They were brought to a concentration camp in 1942 and killed there.

To his friend Max Eitingdon he wrote in 1938:

> *The feeling of triumph on being liberated is too strongly mixed with sorrow for, in spite of everything, I still greatly loved the prison from which I have been released. The enchantment of the new surroundings . . . is blended with discontent caused by little peculiarities of the strange environment; the happy anticipations of a new life are dampened by the question: how long will a fatigued heart be able to accomplish any work? The struggle is not yet over.*

The struggle may not be over as Freud himself said, but what interests me is that Freud, who fought so hard to gain the respect of scientists in his day, is now being taken seriously by a group of neuroscientists. Susan Greenfield argues that his work, so often accused of being unscientific in this century, may now provide a clue to the great question of the next century – the nature of consciousness.

> I am quite unusual, perhaps, as a neuroscientist in finding Freud inspirational. Neuroscience tends to concentrate on the nuts and bolts of the brain, on things you can measure and Freud throughout his life, increasingly so, emphasised the nature of

> the subjective and the whole person. One of the reasons I admire Freud, above and beyond, perhaps, his specific theories, was that he was a pioneer, and I admire anyone that thinks originally and has original ideas. As Einstein once said, 'I bring you new ideas'.

I asked Susan Greenfield for her view of the backlash that there has undoubtedly been against Freud in the second half of this century.

> I think part of the problem nowadays is that scientists especially tend to forget that they do stand on the shoulders of giants. They laugh at people in the past who had misguided views with the glory, of course, of hindsight as they stand by their posh equipment and their button-pressing things and computers. It is so easy to deride seemingly simple-minded ideas of the past. I think one of the reasons that Freud nowadays is not popular among neuroscientists is because of this problem of objectivity, that is one of the main things, and that he was dealing with the mind. The interesting aspect of this is that I think nowadays, certainly among a certain minority of brain researchers in which I include myself, one is beginning to think that if you are going to try and explain consciousness you have to – what is an anathema for most scientists – you have to take on board the

subjective; what a philosopher would call 'qualia'. You have to think about your headache that I cannot hack into, the taste of claret to you that I will never know, the colour of red to you that I can never share.

These aspects of consciousness – the quintessential subjectivity which fascinates philosophers, and indeed would form part of a human being's make-up, would form part of the person that was lying on the couch, laying bare their reactions – those things are an anathema to the scientist. It absolutely goes against all science training to do that. I think that is one of the reasons for the backlash.

Another reason is, of course, they think of Freud as a slightly chauvinistic character. But in fact he was not. He was one of the first people to have evoked derision from others when he said that men could be hysterical and there were smarty-pants people saying 'Well, we all know hysteria is connected to the womb, therefore how can a man be hysterical?' He also wrote about and was interested in, but did not exploit, the issue of transference, that is to say what nowadays would be regarded as, potentially, sexual harassment when his patients started finding him attractive or seeing in him some lover-figure. He wrote very sensibly and clearly about that and he did not take advantage of it, as far as we know. All his writings suggest that he did not, so it might be that people in a feminist mood mistakenly think of him as some lecher with the hysterical female patient on the couch. It was not like that.

Yet he has been a very controversial figure, and in a way has been a unifying hate figure for certain sections of the feminist movement. I asked Susan Greenfield how she reacted to accusations that he was guilty of perpetuating the stereotype of a neurotic, out-of-control, hysterical woman, subject to penis-envy and so on.

I do not think he perpetuated it. The fact is that a lot of his patients were like that. I think that an intelligent woman at the time, if she was not married or was unhappily married – and that would I am sure account for large numbers of them – would have no outlet for her intelligence or her expression, it would not have been focused or channelled. So women, because of the times not because of Freud, may have behaved in this 'stereotyped' hysterical way. Freud met a need. I do not think he perpetuated it. I think his patients reflect the times. One cannot really separate a person totally from their times and I think that a lot of the problem with both scientists and feminists who attack Freud is that they forget the times in which he lived.

Oliver Sacks also perceives a move back towards Freud's

method of studying patients. Freud's artistic gifts found expression in his case histories, which were so elegantly written that they were often read, much to his disgust, as romantic fiction. A great admirer of Freud's writing, Sacks believes that case histories like Freud's, long branded an unscientific indulgence, are coming back into fashion.

> When I was training forty and thirty years ago, I think case histories were already going out and obsolete and regarded as dated, and I think now the pendulum is swinging again. One has seen the re-establishment of case histories such as Freud himself would have loved, and I think there is a renewed sense of his mastery of this form.

I started reading Freud at university almost by accident. I bought some of his books in a sale that a fellow-undergraduate was having to raise some money and when I began reading them, I discovered they were beautifully written. I knew nothing whatsoever about psychoanalysis, but I was drawn into the way he expressed things. The case histories to me were very like stories. Perhaps Freud's power is partly to do with the fact that he writes so well. I asked Oliver Sacks for his opinion.

> I agree. He is a beautiful writer and some of the case histories are like novellas and I think this is one of the things which both attracts people and allows a suspicion that narrative – and vivid, almost

> novelistic narrative – may have to be used if one is going to describe the life of a mind.

But while Susan Greenfield and Oliver Sacks, the neuroscientists, believe Freud's work will contribute to the science of the twenty-first century, Adam Phillips, the analyst, wonders if psychoanalysis, Freud's great brainchild, which has revolutionised our understanding of human behaviour, owes more to the world of letters than to the science laboratory.

> Freud is first and foremost a great writer. I think a bit like your experience, lots of people had the experience of coming across a book by Freud and being very taken by it. It never lets go, once you have read it, it is very hard to forget. That in itself is very interesting. Why cannot we just forget psychoanalysis, why cannot we just lose interest in it? I am sure eventually people will. I think psychoanalysis would not have happened if Freud had not been such a good writer. I think in a way it is that way round. It was almost as though he added science to Shakespeare, Dostoyevsky and Sophocles and produced psychoanalysis.
>
> I think it dawned on him quite early on that in fact the people he was really competing with were Dostoyevsky, Shakespeare and Sophocles. You know he says in the Dostoyevsky paper about how, faced with the artist, the psychoanalyst lays down

> his arms, as though there is a fight going on here and you wonder 'Well, what is the competition for, exactly?' And I think Freud must have known that he was part of a larger cultural conversation and that psychoanalysis may be more of an art than a science.

Which gives us, I think, an appropriately ambivalent ending to this examination of Freud, currently hovering – according to which authority you accept – between art and science and between the mind and the body.

Marie Curie

(1867–1934)

1867 Born Manya Skłodowska in Warsaw, Poland.

1891 Leaves Poland to study physics at the Sorbonne, Paris.

1893 Graduates first in her class.

1895 Marries Pierre Curie. They have two daughters.

1896 Shows that radioactivity is an atomic property of uranium.

1898 Discovers and names polonium and radium.

1903 Awarded Nobel Prize for Physics with Pierre Curie and Henri Becquerel.

1906 Death of Pierre Curie.

1910 Publishes her fundamental treatise on radioactivity.

1911 Awarded Nobel Prize for Chemistry.

1914–18 Organises X-ray Units during World War 1.

1918 Becomes Director of the Radium Institute in Paris.

1921 Visits America.

1922 Becomes a member of the Academy of Medicine.

1934 Dies from leukaemia in France.

A Woman's Place is in the Lab

VERY EARLY one cold November morning in 1891 a solitary female figure stepped out of the Warsaw-Paris Express at the bustling Gare du Nord in Paris. She must have been exhausted after travelling for three days and three nights with her luggage piled round her, trying to keep warm in a fourth-class carriage. Twenty-three-year-old Manya Skłodowska, soon to be known to the world as Marie Curie, arrived in Paris when it had a fair claim to be the scientific and cultural capital of the world. With only a smattering of schoolgirl French, she was about to register at the Sorbonne. She would be one of just a handful of women studying chemistry with some of the greatest scientific minds of the day.

Manya Skłodowska was born in 1867 in Poland, then dominated by Russia, one of a family of four girls and a boy. Their father had taken part in the Polish uprising against Russian control in 1831, but later gave up active resistance and followed the ideals of the Polish Positivist movement, which advocated education and hard work as more effective weapons. Their mother, who shared his views and was also an ardent Catholic, died of T.B. when Marie was eleven, only two years after the death from typhoid of Marie's sister Zosia.

Susan Quinn, journalist and biographer of Marie Curie, is in no doubt about the influence this upbringing had on Curie's development as a scientist.

> I think that her Polish background was absolutely essential to her being able to do what she did. Her parents were of the intelligentsia, but they came from the landed gentry and they were both teachers, but most importantly they were absolutely committed to the cause of Polish nationalism at a time of tremendous oppression under the Russian Tsars. So I think that she grew up with this kind of feistiness and feeling of resisting the status quo, resisting authority, which served her extremely well when she arrived in Paris, a place which was very much dominated – particularly the university – by the male authorities, where she was definitely in a minority. The fighting spirit, which she definitely had, would not have been acquired if she had been a French, bourgeois, young woman.

Marie Curie was obviously very clever, but I wondered whether she had been clever from the beginning or whether she had worked particularly hard or had been especially well taught. Susan Quinn explains that it was a combination of factors.

> I think both well taught and extremely clever. She was the youngest of five children, all of them were extremely good in school, but they all agreed that she was the smartest. There were stories about her being able to memorise poems on the spot: hear a poem once, go into another room, come back and recite it verbatim, that sort of thing. So she had a wonderful mind right from the beginning. And then she had these parents who were absolutely indefatigable educators. They never stopped, the father in particular. So she graduated with a gold medal from her gymnasium in Poland and then had no place to go in Poland because, of course, the university was not open to women.

So she went West, to Paris, following her sister Bronya who had left Poland to study in Paris some years earlier. What would Paris have been like at the end of the nineteenth century for a young student? Susan Quinn describes her early days there.

> Well, she began by settling in with her sister Bronya in 'Little Poland', which was on the outskirts of town, a Polish community within Paris. But she quickly left that for various *chambres de bonne*, around the Sorbonne. And that was really highly unusual behaviour for a young woman of that time. She was twenty-three; she was living alone in various garret apartments and most of

what you read about French women of the time is that this was just not done. So she was unusual.

I was curious to know how she managed financially, having understood that, although the family came from the landed gentry and they were teachers, they had little money.

Well, it is remarkable that she managed to pay her way. She and Bronya worked out this scheme: her older sister Bronya came to Paris first to study medicine, and while Bronya was studying medicine Marie worked as a governess in the countryside and saved her money and supported Bronya, and then they turned it around, and Marie came to Paris and Bronya supported her. But she lived on practically nothing and of course the Sorbonne was free, so she managed, but with difficulty.

Was she absolutely determined to be a student of science, I asked Susan Quinn?

Yes. I think again because of her Polish upbringing. She was part of a movement in Poland called Polish positivism. The idea was that through learning, through science, we can overcome oppression, and I think she was deeply imbued with that. I think she never wavered from that passion for it.

Marie Curie's passion is evident in this passage from her journal, written at that time in the early 1890s.

> *All my mind was centred on my studies. I divided my time between courses, experimental work and study in the library. In the evening I worked in my room, sometimes very late into the night. All that I saw and learned that was new delighted me, it was like a new world opened to me, the world of science which I was at last permitted to know.*

One of only two hundred and ten women amongst nine thousand men at the Sorbonne, Marie had to battle against the prevailing stereotype of the female student as depicted in a popular journal of the day:

> *What distinguishes the serious female student, nearly always a foreigner, is that almost no one takes her seriously. If she is treated with a certain courtesy, she should consider herself lucky. The jokes that are made about her are not always in the best of taste. The female students work with great patience, as though they were doing embroidery. Their study makes them ugly, they usually look like schoolteachers and wear glasses. In the examinations they recite with admirable exactitude what they have learned. They do not always understand it.*

Marie Curie's astounding achievements were part of what began to change that rigid and debarring stereotype. After only two years of studying physics under the tuition of some of the greatest contemporary teachers, Marie Curie graduated first in her class, despite the iron prejudices.

Françoise Balibar, Professor of Physics at Paris University, remarks on how she would have been perceived.

> It is complicated because they did not think of her as a thinker, but they did not think of her as a woman either. Because a woman must not think and a thinker must not be a woman!

Admiration for her resolution increases strongly in considering these years at the Sorbonne. Susan Quinn appreciates what Marie Curie got out of her time at the prime university in France.

> It was a very exciting time at the Sorbonne, particularly in science. Henri Poincaré was there, along with a number of others whose names we continue to celebrate. Lippmann, who discovered colour photography, was one of her teachers. He was one of her great advocates and supported her later on when she applied for membership of the Academy.

Shortly after graduating in 1895, she married Pierre Curie, who was already a renowned physicist when they met. I asked Susan Quinn whether this was a great love story or a meeting of minds, companionability or science? Or both?

> It began as a meeting of minds. They were introduced because Pierre had laboratory space and

> Manya Skłodowska, as she was then, was looking for laboratory space, but I think it quickly became a great love affair. They did complement each other. He was more a dreamer, he was more playful, I think. I think he kind of lightened some of her seriousness, helped her to play, to love nature, to have fun.

In a letter very early in their marriage, she wrote:

> *I have the best husband one could dream of. I could never have imagined finding one like him. He is a true gift of heaven and the more we live together the more we love each other.*

Susan Quinn is almost equally enthusiastic about Pierre Curie.

> Pierre was very much an outsider, he did not play the game, he would not accept the *Légion d'Honneur*, he would not kowtow to the powers that be at the Sorbonne. Now people are re-examining a lot of Pierre's work and much of it turns out to be very important. The work on symmetry in crystals, for instance, is continuing to have significance and resonance. They complemented each other. Marie wanted results, she wanted recognition. Pierre once said that it does not matter who gets the credit as long as

> the discovery is made. She pushed Pierre to finish things and to follow through, for example, the isolation of radium. Their daughter said that, left to Pierre, he would not have bothered about it.

Professor Dominique Pestre is Director of Research at the National Scientific Research Centre in Paris, where I spoke to him about the Curies and their vital professional relationship.

> Pierre Curie of course introduced her to a very specific milieu of French intellectuals. It was a group mainly of scientists and one which had a very strong unity in the sense that they were quite often together as they lived in Paris in the same sector – sometimes in adjacent houses. On Monday afternoons, for example, in the laboratory, you had a kind of gathering of Parisian intellectuals, but it might be novelists, it might be politicians, that was a place where people met, built relations, discussed the future of socialism, of France – things like that.

Another popular topic of discussion was the Dreyfus Affair. A victim of anti-Semitism, the Jewish army officer Alfred Dreyfus had been falsely accused of treason and imprisoned on the notorious Devil's Island in South America. His case had became a *cause célèbre*, splitting French opinion in two. The Curies and their left-leaning friends petitioned his cause at every opportunity.

Marie was working as hard as ever. She had gained a second degree in mathematics, taken on her first job and given birth to her first child, Irène, in 1897. She now turned her thoughts to her doctorate and chose to take up the work of French physicist Henri Becquerel. Susan Quinn takes up the story.

> Marie was looking for a subject for her PhD thesis, her dissertation, and Becquerel had already done this work with uranium. He had found that uranium was giving off strange rays, he called them Becquerel's rays. He did not understand them but he knew there was something going on that was intriguing. He kind of dropped the ball and that research was there for someone to take up. Most of the French establishment was excited about X-rays at the time. This was a kind of side track. So I think with Pierre's guidance, Marie picked it up and started to examine uranium and, because she was a very thorough person, she gathered up every possible element and compound to test.

It is always exciting to know if possible precisely *when* something is discovered. Did the Curies, I asked Susan Quinn, know that they were on to something important and how could they tell?

> It happened very quickly. Marie, in her scrounging

> around, gathered some pitchblende and used a method they had devised to measure it. She found it was giving off much more energy than uranium. She was astounded by this and did not believe her results at first but, as she repeated them, discovered that in fact this was the case. From that they were able, fairly quickly, to deduce – this all happened within the space of a year – that there was something in this pitchblende, there was another element in this pitchblende. That was very important because they had thereby established that it was possible to identify new elements by measuring the amount of radioactivity they were giving off. There were several elements in the pitchblende, but one of them was radium, which turned out to be enormously radioactive, and that was important because the radioactivity provided clues to the nature of matter and really opened the door of the nuclear age.

The prescience of scientists is sometimes acclaimed but, as often, its lack is criticised. It would have taken a bold leap of imagination, I thought, to see the nuclear age. Did they sense it at the time?

> They did. Marie, in her second paper, described this as an atomic phenomenon and that is remarkable, very, very prescient because, at the time, people were still debating about whether

> matter was made up of atoms, and here she was saying that this was an atomic phenomenon, and it did not take very long for other people, particularly Ernest Rutherford in England, to figure out that, if you could play around with this radioactive emission and separate out the various elements in it, you could perhaps understand the structure of the atom.

I asked Susan Quinn how Marie Curie's work had been furthered in England.

> What was happening in England was that J. J. Thomson had already discovered the electron, so that component of the atom was beginning to be understood. Rutherford began to work with radioactivity. He had very little in the way of radioactive material to work with, and much of it came from Marie Curie's laboratory, in fact, but he did continue to work with it and he began to understand that there were several kinds of emissions. What he concluded – and this is Rutherford's great discovery – was that, in the radioactive process, heavy atoms were disintegrating and you could then separate out these disintegration products and understand the structure of the atom.

Dr John Gribbin, broadcaster and author, explained to me the astonishing significance of this step forward.

The great revelation in the work that Marie Curie did on radium was twofold, really. First of all she showed that atoms are not indivisible, that there are things that come out of atoms: these mysterious rays are, in fact, particles which are being emitted by atoms, and in the process it turns out atoms are turning into something else. So we are getting this idea of the atom, or the nucleus as we would say today, as no longer being indivisible but as being something that can be changed into something else. That gives us a whole idea of modern versions of radioactivity and eventually atomic fission and so on.

But the thing that really happened in this work, and what paved the way for other researchers like Ernest Rutherford, was using the stuff as a tool, and this is one of the most fascinating aspects of the story. Here is somebody (Curie) who comes along, discovers the radiation, starts to work on one aspect of the problem, sorting out a new element, making a new discovery which was dramatic in its time, but then the radiation itself is very, very important. What happens in radioactive material is that the nucleus of the atom spits out this particle, called an alpha particle and, in the process, it turns into something else, it turns into another element: it is a transmutation of the elements, the alchemists' dream. And then that alpha particle, being a very fast moving, hard little particle, like a little bullet, can be used as a detector to fire into other atoms and see how it bounces off them, ultimately to split them. It is by firing particles at unstable atoms that you make the nuclei split apart and you cause fission.

With hindsight you could say that Marie Curie was almost the mother of the atomic bomb. The work that she did follows in a direct line of descent through Rutherford to the splitting of the atom and then to the atomic bomb project itself. It was through that beginning of modern particle physics, firing things into things to see what comes out the other side, that really most of twentieth century physics develops.

The isolation of radium and the discovery of radioactivity is a story familiar to us all. It was an achievement of enormous significance and one that Marie Curie had worked on tirelessly day and night, manipulating massive quantities of noxious substances in primitive conditions. This is the storybook image of Marie Curie that I remember from my schooldays, a diminutive figure stirring bubbling concoctions in a vast cauldron. The stress, however, of such hard work took its toll. Marie suffered a miscarriage before giving birth in 1905 to another baby girl, Eve. Pierre abandoned his own research to help her in the laboratory and the couple collaborated on the work. But it was Marie's tenacity that kept the project going over the years when the result seemed a long way off. Susan

Quinn is convinced that this tenacity was at the root of her achievement.

> Marie was somebody who wanted results and, whatever denials she made to the contrary, she wanted to succeed and she wanted recognition. Marie knew that if they did not isolate the element they were not going to be credible and she pushed Pierre.

Dominique Pestre agrees with this verdict.

> If someone persisted in searching for radium, it was Marie Curie, not Pierre Curie. For a while Pierre Curie withdrew and it is only after a while, after Marie Curie had shown how important it was to do that job, that Pierre Curie came back.

Exhausted but invigorated by their work, the couple proudly exhibited their lumps of luminous material in their under-equipped laboratory, as yet unaware of its lethal potential. The scientific establishment had to sit up and take notice of the Curies' work. In 1903 the Nobel Committee awarded Pierre and Marie, along with Henri Becquerel, the Nobel Prize for Physics for their research into radioactivity. Susan Quinn sees this as a most significant turning point.

> I think it elevated the Nobel Prize to new fame

> and celebrity in the world. That was only the third year the prize was given and people were absolutely electrified and astonished by the fact of a scientific couple and there was a lot of sort of romantic gibberish about this wonderful couple and love in the chemistry laboratory and so on, and they were hounded by the press. They became celebrities, much to their chagrin. They really hated it.
>
> But the Nobel Prize also helped them. They complained a lot about the way in which it changed their life for the worse because they became so famous. But as a result of the Nobel Prize, Pierre was admitted to the *Académie des Sciences*. That is important, because it means you get more funding in France. He was made a Professor of the Sorbonne. He got to move over from the *Ecole Polytechnique* to the Sorbonne. They got funds for their laboratory. That was the most important thing, although Marie Curie was never accepted by the *Académie des Sciences* – and she was very hurt by it. She was marginalised by that, but it steeled her to the purpose of establishing her own permanent institute. She went out and proved herself on her own.

Marie Curie may have helped put the Nobel Prize on the map, but it seemed difficult for anyone at the time to acknowledge her true part in the discovery. The press both in France and abroad usually cast her in the supporting role with comments from the *New York Herald* like '*Mrs Curie is a devoted fellow labourer in her husband's researches and has associated her name with his discoveries*' and '*Monsieur Pierre Curie is ably seconded by his wife*' from *Vanity Fair*. Dominique Pestre takes up this theme.

> All the letters about radioactivity sent to the Curies before the death of Pierre Curie, all of them in the archives were addressed to Pierre Curie and never to Marie Curie, which is quite interesting. If you take the Nobel Prize, the first proposal was for Becquerel and Pierre Curie to get it, not Marie Curie, and it was Pierre who specifically asked to have Marie Curie included. She was not in general considered as a genius at all. She was even described in private letters – not sent to Marie Curie but between men of that group – as a not particularly imaginative and good scientist. So if we put all that together we have the image that Marie Curie was a kind of second-rank person in all that affair and that particular point we know is false.

Despite her lack of acknowledgment, Marie Curie continued to work with her husband. There seems to have been no professional rift between them and no evidence that they lessened in their urgency. Then, tragedy struck.

Every day, on his way to work, Pierre Curie walked along the Quai des Grands Augustins on the Left Bank towards Pont Neuf. One morning, with his thoughts elsewhere, at the point where the Pont Neuf meets rue Dauphine, he stepped in front of a horse-drawn carriage and was killed instantly. His skull was crushed. Marie was distraught. Her beloved husband and her closest colleague had gone. Susan Quinn has no doubt of its dramatic effect on Marie Curie.

> It was a devastating event in the life of many people who loved him but particularly Marie Curie. In the year after his death she kept a journal and addressed him in the journal as though he were still alive. She described her feelings and her wish almost to walk out into the street herself and be run over. And the depth of her mourning is really powerful. She also, in this diary, describes their life together just before he died. She recapitulates the last hours that they had together before he died. That is a tremendously moving document and convinces me of her really profound love for him. She writes:
>
> *I enter the room. Someone says 'He is dead'. Can one comprehend such words? He, whom I had seen leave looking fine this morning, he whom I expected to press in my arms this evening, I will only see him dead. And it is over for ever. I repeat your name again and always, Pierre, Pierre, Pierre, my Pierre. Alas, that does not make him come back, he is gone for ever, leaving me nothing but desolation and despair.*

> She stopped working. She did not go to the laboratory for a long time. She became depressed.

I asked Susan Quinn if Marie Curie blamed herself in any way, as people sometimes do.

> Yes. It is interesting, there are a few pages torn out of her journal. I suspect that those are pages where she talks about the difficulties in the relationship and in the marriage. That is just a guess. There is one hint that I think is very powerful. She thinks about the moment that he left the house for the last time. She is upstairs with the two young girls, their children; he is downstairs at the door, and he says to her 'Are you coming to the lab, Marie?' and she says 'Don't torment me'. And those are her last words to him. And she feels terrible that they were those words and not words of love. But I think it is very poignant, because you see this woman with young children who is pulled in several directions: towards Pierre, towards the laboratory and towards the children at the same time.

But the popular image of Marie Curie seeing out the rest of her days in black is false. Letters, written several years after Pierre's death, revealed her in the throes of a passionate affair with a married man, Paul Langevin, a physicist. France was scandalised. Leaked to the press by an insanely jealous wife, these letters were splashed across the front pages of the popular press, who vilified Marie as a husband-stealing foreigner, no longer welcome in France. Susan Quinn elaborates.

> The relationship began because Marie was grieving for the loss of Pierre and Paul Langevin was miserably unhappy in his marriage and trying to figure out what to do, and so they consoled each other. That turned into a love for each other, a real, passionate love affair. Paul Langevin had been her husband Pierre's best friend and had written a very beautiful eulogy to Pierre at the time of his death, so I think that she felt connected to Pierre through Paul Langevin. It is a very understandable thing, two very lonely and unhappy people who found each other.

In the middle of this outrage – as it became – the Nobel Committee awarded Madame Curie a second prize, this time for chemistry, in honour of the discovery of the two radioactive elements, radium and polonium. However, when the Committee got wind of the scandal, they wrote and suggested she should not come to claim her prize until she had cleared her name. They thought it would embarrass the King of Sweden. Marie Curie, Susan Quinn reports, was incensed.

> I feel it really reveals her at perhaps her finest

> hour. She wrote back and said she had understood that the Prize was being given to her for her scientific work, that her private life had nothing to do with it, and that she was coming to Sweden to accept the Prize. And she did. I think it was one of the hardest things she ever did.

If she hoped that her extraordinary achievement in winning a second Nobel Prize might silence her enemies in France, she was to be disappointed. Susan Quinn comments.

> Surprisingly, I think, many of the French continued to see her as a wanton woman. Even now you can hear people in France talk about Marie Curie as a kind of loose, fast woman. I was really surprised that there was an undercurrent of rumour still about her. So I do not think the bourgeoisie ever got over that scandal. I think that the affair and the scandal and humiliation really motivated a lot of what she did in her life after that. For instance, she became very involved in mobilising X-ray units – X-ray mobiles they were called – to go to the Front during World War One to X-ray wounded soldiers. And I think she was a very patriotic person anyway, but I think she was also motivated by the need to rehabilitate her reputation after that affair.

Support for Marie Curie's cause did rally from various

quarters including her family, Pierre's family and some in the scientific community. Einstein was rude about her personally (*She is not attractive enough to become dangerous for anyone*, he wrote to one friend), but defended her with a passion and wrote her supportive letters when the scandal was at its height.

> *I feel the need to tell you how much I have come to admire your spirit, your energy and your honesty. I consider myself fortunate to have made your personal acquaintance in Brussels. I will always be grateful that we have among us people like you – as well as Langevin – genuine human beings, in whose company one can rejoice. If the rabble continues to be occupied with you, simply stop reading that drivel. Leave it to the vipers it was fabricated for.*

In 1921, Marie Curie made a publicity trip to the United States with her two daughters in search of funds for her work. She allowed herself to be promoted as an impoverished, humble widow, martyring herself to the cause of medicine, and thus the myth of Marie Curie was born. She was described as the saviour of mankind who was going to find a cure for cancer, although Marie Curie never claimed this. She returned to France the proud owner of one precious gram of radium, then worth $100,000.

The irony was, of course, that over-exposure to her radioactive materials was killing Marie Curie. Her adamant assumption that scientific progress meant a move towards an ever-improving world prevented her and many others from perceiving the increasingly obvious dangers of what she was dealing with. At her laboratory in the rue Pierre and Marie Curie, which she set up in 1914, her notebooks are still held under lock and key, too radioactive to be touched by human hands. Susan Quinn describes how the realisation of the effects of radioactivity came too late.

> She gradually came to understand the dangers. She died of leukaemia in 1934, which was a result of her exposure to radioactivity over many years. In the early days they did not understand at all what was going on. They knew that the radioactive materials would burn the skin, but they did not understand that there was systemic damage to cells going on. That was really not until the 1920s. There were women who were painting watch dials with radium and they were tipping their brushes with their tongues. They developed carcinoma of the jaw, many of them died, and it was as a result of that incident that the dangers of radioactivity began to be taken very seriously.

Whenever Marie Curie was asked, in her later years, if she was going to write her autobiography, she responded with what I think is a touch of false modesty by saying that her life was such an uneventful, simple little story. There is no doubting the dangers of what she did, nor that she overcame enormous prejudice as a woman and a foreigner.

But how scientifically significant is her work in radioactivity? Exactly a hundred years on from the discovery of radium, John Gribbin assesses the lasting legacy of her discoveries for science today.

> It is very tempting to say that the Curies' work opens the way for nuclear physics, that it is the dawn of the atomic age, the nuclear age, and of course it all happened at the beginning of a new century, so it is nice to look at it in those terms. You can certainly make that case. I do not like to think, though, that it would not have happened if they had not done it. It was something that was going to happen, these ideas were around, people were discovering radiation, there were people like Rutherford around and I do not subscribe to the view that science is the result of some unique genius who comes along and makes a breakthrough.
>
> I think the time was right for that work, she was the right person in the right place at the right time and she did a superb job and probably hastened the progress of science, but it would have happened anyway.
>
> The medical side is probably the biggest influence that the Curies have had – certainly on everyday life in the twentieth century. There are two sides to radiographic work. There is the basic X-raying, which is absolutely crucial and which was the thing that was so important in her work during the First

World War, finding where the bullets were in people that had been wounded, things as simple as that. But then there is the other development involving radium which comes right back to her own discoveries, which is treating cancer cells, where you use the alpha particles, as we now know they are, to kill off the dangerous cancerous tissues in a cancer.

Leaving aside the medical applications which are obviously so important, if you are a scientist today and you work with one of the big accelerators, you think of these huge machines which accelerate particles to absolutely enormous amounts of energy and do all these amazing and fantastic things with them, but right down at the bottom, at the beginning of that whole machinery, you have got to have some particles to accelerate and those particles that you are starting out with, in some cases, are still the same particles that people like Marie Curie and Ernest Rutherford worked with, and they come from the same sources, they come from natural radioactive decay of substances like radium.

Now instead of just letting them come out of the radium and hit into an atom, you grab hold of them with magnetic fields and you whizz them round in accelerators to huge energies and then you let them out the other side and you do all the interesting things you do with them. But right down at the base level you could not do it if you

> did not have the particles to start with and you have those particles because we know what the Curies discovered.

Susan Quinn takes another view.

> I would say personally what I most admire and feel strongly about with her, is her courage. She was a person of tremendous courage. I think the story of the Nobel Prize is an example of that but there are many, many times in her life when she did very daring things. I think that her importance is growing. You may know that recently France decided to move the ashes of Marie and Pierre Curie to the Pantheon. She became the very first woman to be buried there because of her own accomplishments. So I think that there is more recognition of her work in France and I think the anniversary of this discovery may cause us to re-examine the enormous significance of her discoveries.

But John Gribbin's conclusion is different. He says we can take this reassessment of the most famous of women scientists too far.

> There are very, very few great scientists. Everybody will agree on Newton, and then you say 'Well, Einstein' and then you start saying 'Er-um-who?'

I think the great scientists can be counted on less than the fingers of one hand. Marie Curie was a very important scientist, which is not quite the same thing. The history of her story is almost as interesting as the story itself. She was hailed, first of all, as being an absolute genius because she was a woman who had achieved all these great things, and I think there was a feeling that, in order to achieve anything, you must be a double genius if you are a woman because most women have not achieved it. So there was a generation that was brought up on Marie Curie the great genius, indeed to rank with people like Einstein and Newton.

Then I think you got the backlash, people coming along saying, 'Oh come on, she was not that good. You know, anybody competent could have done that kind of work at the time, she was just lucky', and I think that goes too far to the other extreme. Perhaps now, at last, more or less a hundred years after the work was done, we are starting to see a real balance, the pendulum has swung to both extremes and it is settling down a bit in the middle. She was a really good scientist, but not of the very first rank. But I wish I were as good as her.

A characteristic of major scientific discovery seems to be that the significance of the work can often change, and radically, as time goes on. To those who know Darwin well, for example, his work is becoming increasingly important and the more that is discovered around it the more central it seems. Perhaps in time Marie Curie's work will be seen in a perspective which indisputably raises her into what John Gribbin calls 'the very first rank'. Susan Quinn believes that she is there already – as, of course, do other contributors to this book, who would not pause after Einstein and Newton and would certainly propose Archimedes, Galileo, Darwin, Faraday and so on, and be more inclined to anxiety that too many great scientists – Copernicus, Kepler, Maxwell, Penrose – had been omitted.

Albert Einstein

(1879–1955)

1879 Born in Ulm, Germany.

1896 Renounces German citizenship. Enters the Swiss Federal Institute of Technology.

1901 Takes Swiss nationality. Publishes first papers on forces between molecules.

1902 Moves to Bern to take up job as Patent Officer. Birth of illegitimate daughter, Lieserl.

1903 Marries Mileva Maric. They have another two children, both sons.

1905 Publishes 'On the Electrodynamics of Moving Bodies', introducing Einstein's special theory of relativity, as well as four other papers on Brownian motion and the photoelectric effect.

1909 Appointed Junior Professor at the University of Zurich.

1910 Becomes Professor of Theoretical Physics at University of Prague.

1914 Appointed Director of Physics at Kaiser Wilhelm Institute, Berlin.

1916 Publishes *The Foundation of the General Theory of Relativity*.

1919 Eddington's report on the solar eclipse proves Einstein right. Divorces Mileva and marries his cousin Elsa.

1921 Awarded the Nobel Prize for Physics.

1932 Leaves Germany on visit to America and never returns.

1934 Becomes Professor in the Institute of Advanced Study at University of Princeton.

1952 Invited to become President of Israel.

1955 Dies.

The First Celebrity Scientist

FOR MANY of us the very word 'genius' conjures up an image of Einstein in his later years – white-haired, wild-eyed, and wearing no socks. He is probably the most widely referred to scientist of all time and possibly the one whose work is least known in any detail. Because he lived so close to our own times and left so many records, we can look more closely than before at this troubled word 'genius'. I asked Professor Sir Roger Penrose, Rouse Ball Professor of Mathematics at Oxford University, whether he considered Einstein merited the term.

> If the word 'genius' should be applied to anybody, then Einstein is one of the people to whom it should be applied, particularly with regard to his discovery of general relativity.
>
> It is one of those theories which might not have been arrived at by anyone else. Often one thinks that there is this relentless march of science. It does not matter much who gets there. But I think

> Einstein's general relativity is an example where it was not part of this march of science. It was something quite outside the way in which people had been looking at things. It was a highly original idea and a very profound one, and I could quite believe that it might not have developed even by now.

Albert Einstein was born in Ulm in Germany in 1879, into a closely knit Jewish family. His father was an unsuccessful businessman who uprooted the family many times around Germany and Italy, where they finally settled in Milan when Albert was fifteen. Yet it seems to have been a comfortable, happy childhood. John Gribbin, who has written about Einstein's life with Michael White in *Einstein, A Life in Science*, describes it.

> There was a tradition at that time that Jewish families would take in for a meal once a week a poor Jewish student, and they had a student who used to come and eat with them who was studying science and would talk to Albert about science and what was going on.
>
> Einstein's later reputation for a certain slowness in his early years came from the fact that he was lazy at things he was not interested in. He hated the very rigid school system in Germany at that time, where you would literally get a rap over the knuckles for a wrong

> answer. But he was interested in things that interested him.
>
> There is a famous story about him being given a compass when he was a small boy when he was too ill to amuse himself, and his being fascinated by the way the compass always pointed north, and worrying about that. He was deeply interested in all of that stuff and then he had to go to school and learn things by rote and he just switched off. I do not think he was ever lazy in the sense of could not be bothered about anything. He just hated the system. And this applied when he went to university. Even when he was doing his first degree, he just did not bother going to lectures. He went to things that he was interested in and spent all the time in the library reading things up for himself and when it became time to do the exams he had to borrow a friend's notes and mug up on them and just scrape through the exams as a result.

With his poor marks, it is hardly surprising that Einstein found it hard to get a job at a university, but some friends helped him find a not very taxing one at the Swiss Patent Office in Bern. In the mornings Einstein worked on patents, but in the afternoons, secreting his books under his desk, he began to formulate the theories which would revolutionise science. He published five important papers in 1905, while he was still an unknown at the Patent Office. In one of them he fearlessly took on the single most important issue then facing physicists: how to reconcile the contradictory theories of Sir Isaac Newton and Sir James Clerk Maxwell. John Gribbin outlines the issue.

> The problem people had at the end of the nineteenth century was that you had two sets of fundamental laws in science. You had Newton's laws of physics that had been around for more than two hundred years, which everybody regarded as the ultimate truth, and you had Maxwell's laws which described electro magnetism and light and everything to do with electricity and magnetism, and they were new but they seemed to be fundamental. They disagreed with one another and almost everybody said, 'There must be something wrong with Maxwell's ideas because they are the new ones. Newton must be right.' Einstein's genius was to say 'Let us consider the possibility that Newton is wrong', and when he did that he found you could make a match between his version of mechanics and Maxwell's equations of light.

Einstein began by rejecting the notion of the ether. According to scientists at the end of the nineteenth century, this was a mysterious substance that filled all space, and through which light, electricity and magnetic forces travelled. Einstein thought this highly unlikely.

His most striking contribution to science came not from work in the laboratory but from what he called 'thought experiments'. The special theory of relativity, published in 1905 in his paper 'On The Electrodynamics of Moving Bodies', came from one such brilliant idea. Einstein imagined what it would be like to ride a wave of light at the speed of light. Realising that James Clerk Maxwell could be right and the speed of light might indeed always be the same, Einstein introduced the extraordinary idea that time might vary instead.

Einstein's influential contribution concerned the nature of space and time. Paul Davies develops this aspect of Einstein's thought.

> Newton had introduced space and time as parameters, but they were just an arena in which the great drama of nature was acted out. For Einstein, space and time became part of the cast: that is they became dynamic entities subject to change, indeed manipulation. We can manipulate time in the laboratory. It is not a fixed thing that is there for everybody. It is not a universal and absolute background against which things happen.
>
> The notion that time itself is part of the physical universe was entirely novel. Now these are not just theoretical effects, they are measurable. If you put a clock in an aircraft and fly it around for a few hours and then compare it with its clone on the ground, there is a measurable mismatch, albeit only a few billionths of a second, but well within the capabilities of atomic clocks.

> We know that space and time are things which can be changed and which depend on your state of motion and depend upon your gravitational circumstances. I think that is probably the most important departure from Newton's world view and of course it leads to all sorts of consequences about the motion of bodies. Most famously it leads to the conclusion that nothing can travel faster than light. But I guess the most dramatic manifestation and the difference between Newton's physics and Einstein's physics concerns the only equation your average man and woman in the street actually knows from physics: $e=mc^2$.
>
> Einstein showed that space and time are linked together and he also showed that energy and mass are linked together. This leads on to nuclear power and nuclear bombs and things like that. This encapsulation of energy and mass is manifested most obviously in the sun. You look up at the sun and it is shining. What makes the sun shine is the conversion of some of its mass into the energy of heat and light, because it is basically a nuclear reactor. So it does lead to observable consequences, consequences which are actually quite crucial in everyday life.

The general theory of relativity (published as 'The Foundation of the General Theory of Relativity' in 1916), Einstein's greatest achievement, came from another thought experiment which he had while sitting at his desk in the Patent Office – what he called 'the happiest thought of my life'. He imagined someone falling inside a lift which had broken its cables. Just like an astronaut in space they would free fall inside the lift. John Gribbin elaborates.

> The general theory is a different kind of theory because it is about accelerations, but the insight Einstein had was that acceleration is exactly the same as gravity. Through that he developed this idea of bending space and time; bent, space-time enables you to deal with things like orbits, not as a force tugging on a planet from the sun mysteriously, across space, but as the planet rolling around a dent in space, rolling around the sun like a marble rolling around in a round bowl. That gives you a completely different view of how the universe works. The thing that I am most fascinated by in all this is that it tells you how the universe began and how it evolved because it is the general theory of relativity that gives you all this business about black holes and the Big Bang and cosmology, which is all wonderful, wonderful stuff and of course completely useless. I mean, no one has ever found a practical use for the general theory of relativity yet.

Einstein's general theory of relativity does not have much impact on our everyday lives, yet it seems to have completely changed the way we look at the universe. This seems something of a paradox. Perhaps there are two worlds – the abstract external world out there and our own concrete world – which meet only now and then. John Gribbin takes up the point.

> It is true that it has not had much direct impact. There are things such as these devices which can locate your position to a very great precision which actually use Einstein's general relativity, the very accurate ones do and, in space travel, the precision that one now needs does use Einstein's theory, but you are right, it is mainly a change in outlook. It is not as though it has directly influenced our lives in any particular way. I think that is true.

Later Einstein recalled a conversation with the great physicist Max Planck, who was on a visit to Zurich. When Einstein told him what he was working on, Planck said 'As an older friend, I must advise you against it, for in the first place you will not succeed and, even if you succeed, no one will believe you'.

Scientists at the time and ever since have found the general theory of relativity extraordinarily difficult to understand, as even Professor Stephen Hawking admits. In fact Einstein himself is supposed to have told his publisher that his popular book on the subject would be understood by only twelve people in the whole world. But the difficulty of Einstein's great insight did not stop people recognising the importance of what he had discovered.

Scientists knew that an eclipse of the sun would prove whether Einstein was right or not. Various international expeditions had set out to test it, but in each case the expeditions had been turned back because of the First World War. It was not until after the War finished, in May 1919, that two British expeditions reached Brazil and West Africa to measure if light was deflected as Einstein had predicted. They sent back joyous confirmation that Einstein was right and the world acknowledged the fact that it lived in a curved universe. Einstein's immortality was assured. On November 7th, *The Times* reported:

> *Yesterday afternoon, in the rooms of the Royal Society at a joint session of the Royal and Astronomical Societies the results obtained by British observers of the total solar eclipse of May 29th were discussed. It was generally accepted that the observations were decisive in verifying the prediction of the famous physicist Einstein, stated by the President of the Royal Society as being "the most remarkable scientific event since the discovery of the predicted existence of the planet Neptune".*

Einstein found himself famous, and his fame never abated.

But there is a darker side to the story of the twentieth century's greatest scientist, including a missing daughter and a mistreated wife. Robert Schulmann, Professor of History at Boston University and joint editor of *The Collected Papers of Albert Einstein*, is the man who uncovered previously missing letters written by Einstein, which shed an unflattering light on his personal life.

> He is very manipulative in human relations. Clearly one should not judge Einstein, the kindly old gentleman who smiles a lot at children, and confuse him with the scientist on the make in the earlier part of his career, certainly up to general relativity, at which point he was not quite forty years' old. This same kind of ruthless opportunism that he exhibited in his science he also exhibited in his personal relations.

Until Robert Schulmann discovered otherwise, Einstein's history had seemed conventional enough. In 1903 he married a Serbian woman, Mileva Maric, a fellow student at the University of Zurich. She did not complete her studies, and the letters reveal one of the reasons why. In 1901, Mileva had become pregnant with Einstein's child while they were still at university and two years before they married. She returned to Serbia to give birth to their illegitimate daughter, Lieserl. Einstein wrote an affectionate letter to Mileva from Bern shortly after the birth. But he never actually saw the child. She vanishes from their lives soon after. In the letter he writes:

> Is she healthy and cries properly? Who gives her milk, is it hungry? I love her so and do not know her yet. I should like to make a Lieserl myself. That must be interesting. It can certainly cry already, but

> will learn to laugh only much later. Therein lies a deep truth.

Robert Schulmann investigated further.

> We do not know what happened to Lieserl the daughter. In the thirties, a woman appeared at Cambridge or Oxford – I believe Oxford – and claimed to be Einstein's daughter. He was informed of this fact by a German friend of his. Einstein pooh-poohs the matter in a letter that is in the archives, but then he turns to his secretary, Helen Dukas, and says 'Find out about this woman'. A private detective is hired. The private detective establishes that the woman is not his daughter and the matter is laid to rest. What is interesting, psychologically of course, is that Einstein, I assume, feared that this was the daughter whom he had abandoned in 1903.

Einstein and Mileva also had two sons after they married, Hans Albert and Eduard, but later the marriage began to fall apart. In 1913, the family left Zurich for Germany when Einstein was appointed Director of Physics at the Kaiser Wilhelm Institute in Berlin, and there he became involved with his cousin Elsa. Robert Schulmann describes Einstein's subsequent behaviour.

> Einstein submits a memo to Mileva in July of 1914. The story is grimmer, I think, than is obvious just

from the memorandum. The memorandum in itself, without any context, is a set of conditions under which he is prepared to continue the marriage and the conditions are very, very harsh – in a sense they reduce her to being his washerwoman. Einstein had originally intended, at the promptings of Elsa, to divorce Mileva in 1914. For one reason or another that we do not know, the divorce never came off. It became a separation.

Mileva returned with the boys to Zurich, Einstein stayed in Berlin and then, in 1916, he made another try at divorcing her, perhaps again at the promptings of Elsa. It is a reasonable conjecture. At that point Mileva had a nervous breakdown. To give one illustration of the callousness of Einstein: he was sure she was faking it and he writes to two friends in Switzerland that she is leading him by the nose, they should not believe this act that she is putting on. Then the friends convince him that she is not faking it and he pulls back from his divorce attempt. In 1917 he is quite exhausted by the General Theory of Relativity and, in the fall of 1917, in order to take care of himself more nicely, he moves in with Elsa, into adjoining apartments. This is at a time when he is still, of course, married to Mileva. And then in January 1918, he makes the third push to divorce and that is the one which a year later he brings to a successful conclusion.

In his divorce settlement, Einstein confidently promised Mileva his share of the Nobel Prize, three years before he was actually to win it.

The second marriage with Elsa seems to have been no happier – perhaps more a marriage of convenience than a meeting of minds. He never pretended to be a family man.

> My aim lies in smoking, but as a result things tend to clog up, I'm afraid. Life too is like smoking, especially marriage . . .

In the twenties and thirties, Einstein became a popular public figure, meeting Charlie Chaplin, corresponding with Sigmund Freud, and staying in the Franklin Room at the White House with the Roosevelts. Thousands of ordinary people wrote to him, including one six-year-old girl who asked him to get his hair cut. He took easily to the role of celebrity scientist and loved to play to an audience. John Gribbin illustrates this.

> He once put down on a form that he had to fill in, under 'Occupation' 'Artist's Model'. Another time he was getting out of a taxi somewhere and someone asked him to pause for a picture and he stopped and they took the picture, and he turned to his companion and said 'Well, the old elephant's done his trick again'. So I think he did enjoy all

> that and I am sure that the business about not wearing socks and so on was to some extent a deliberate affectation to play up to the eccentric scientific image.

In the early 1930s, Einstein left Germany and the Nazis behind for America. He later urged the United States to develop the atom bomb before Germany did the same, although he had no connection with the building of the bombs which were to destroy Hiroshima and Nagasaki. After Hiroshima he said 'If I had known they were going to do this, I would have become a shoemaker'.

Einstein's move into the public sphere with his statements about world peace and the bomb coincided with the decline of his scientific importance. A friend and colleague, Abraham Pais, wrote that after 1925 Einstein should have gone fishing. It was left for the next generation of scientists to take up his theories and prove that they were accurate in the real universe, not just on paper.

Jocelyn Bell Burnell, now a Professor of Physics at the Open University, was part of a small team at Cambridge which discovered pulsar stars in the early 1960s. Pulsars are vital to the history of general relativity because they allow astronomers to test Einstein's predictions with enormous accuracy. Later scientists building on her work realised that these peculiar radio stars came in pairs. She helps to answer the question of how we now know Einstein was right.

> There is a particular pulsar, the very first one that

was discovered to be in one of these twin systems. That one has been studied now for about twenty years. We have studied how that pair of stars moves in towards each other, and it does it exactly in the way that Einstein's theory of general relativity predicted. It is a glorious confirmation of that bit of Einstein's work.

Einstein set up quite a lot of theory – the theory of special relativity, the theory of general relativity – and then basically scarpered off and left other people to prove it. Since then physicists have been pretty busy gradually ticking off one by one the various bits of Einstein's theory. It is standing up very well. It is not yet totally there and therefore it might not be totally correct. It has undoubtedly got intellectual consequences. I would hesitate to say that it has no further effect because one of the things that we have learnt is that even the astronomical discoveries can in the longer term lead to useful spin-offs, but they are usually so far down stream that it is impossible to see them at this stage.

Roger Penrose was drawn to Einstein's work in the 1960s, at a time when physicists had abandoned general relativity and were looking elsewhere, particularly to quantum physics, for answers to major questions about the universe. I asked him why he had chosen to work on a theory that had become so unfashionable.

> I think primarily because it was a theory which just tremendously appealed to me. Part of this is somewhat personal in the sense that I have a very geometrical way of looking at things and general relativity is a very geometrical theory, so it was something that I felt I could come to terms with and understand in a way I would have trouble understanding other theories, which were of more of a directly calculational nature. But I think many people who know anything about the theory would say this is a tremendously beautiful theory. They might find that they would work on something else because that is more practical or more directly likely to lead them to get a job or something. But Einstein's theory has intrinsically this tremendous aesthetic appeal.

Stephen Hawking has written about Einstein's theory of relativity that '*No one can recall without a thrill his first encounter with this Carollian world where space–time is curved, a fourth dimension, and honest witnesses blithely disagree on the most elementary questions of what happened when and where*'. I asked Sir Roger Penrose to explain in layman's terms the impact Einstein's theory had on Hawking, as well as on himself.

> Well, it certainly is an extraordinarily different view of the world from the one that one had before with Newton's theory. I do not think it quite had the effect on me as Hawking describes. When I

first heard of it, it took a little while before I had a properly coherent view of what the theory actually was, but indeed when I did get that view, I did have this kind of thrill which he refers to.

I think that what is so striking about it is that it is simply taking a very simple, rather everyday experience, namely that acceleration feels the same as gravity. If you are in a car and it starts up or stops suddenly, you feel pushed or pulled in the same way as the effect of gravity, and Einstein's ideas develop from an old observation, namely Galileo's, that if you drop objects of different masses they will fall together in a gravitational field. So that if you sit on one and look at the other you would see the other hovering in front of you as though there were no gravitational force at all.

Of course, nowadays we are all very used to this idea because of astronauts going around and the earth is right there but they float and they do not seem to feel any gravitational force, and that is because they and their space capsule and everything are all essentially falling together in the gravitational field. Because of this thing called the principle of equivalence, which Galileo first enunciated, one can get rid of the gravitational field by falling with it.

Another way of putting the same thing is that acceleration feels the same as a gravitational field. And from that simple fact, rather than the kind of picture that Newton had had, with forces acting

> between particles and so on, from the observation of the principle of equivalence, one can develop this completely different way of looking at gravity, where space–time curvature is what is responsible for gravitational effects rather than forces.

I wondered if Roger Penrose felt that he was following directly in Einstein's footsteps, if there was a sense in which he felt a baton had been passed on to him.

> I do not think I felt anything quite like a baton, but I think that was one of the things appealing about Einstein. One did not need to know a lot of other physics. So I think in that sense what you say is true, that Einstein had this theory and one could just take it directly from what Einstein did in 1915, 1916, without worrying about intermediate things that happened.

I thought it worth raising that Roger Penrose has a mathematical background and, from what I have read, Einstein's mathematics were not the strongest part of his armoury.

> Yes, I suppose there are a number of ironies here. That is probably the reason Einstein did not develop greatly after the production of general relativity. People often say 'What did Einstein do after that?' Well, he did a few things of interest but they were

not quite of the momentous character of things he did before that theory.

Einstein, it is said, notably refused to accept an equally important theory: quantum theory. Did Roger Penrose agree with this charge?

I think I should put this in its historical perspective, because quantum mechanics was almost started by Einstein. That is to say he was the second person who really got involved in the theory. Max Planck introduced the initial ideas and Einstein very quickly came in and made all sorts of suggestions about what the nature of that theory must be. They were powerful ideas. Einstein had a very important role in the initial development of quantum mechanics. However, as the theory developed there were particular aspects that he found very disturbing and unsatisfying and not part of what he thought the universe was like – I think primarily the kind of subjective view that the Danish physicist Niels Bohr found himself driven to, which is, in some sense, that at small quantum level it is all in the mind, there is not a reality at that level. Einstein did not accept that at all, he thought there must be a reality even if it is a very strange one down at the quantum level. Personally, I think Einstein was completely right on that. It is just that there were certain aspects of quantum mechanics that he found very hard to accept: the idea that God was playing dice was something that Einstein found very disturbing.

What did he mean by that phrase 'God does not play dice'? I asked John Gribbin about this.

He became very upset by how quantum theory incorporated ideas of probability and uncertainty. What quantum physics tells you is, if you are watching a radioactive atom and you know it is going to decay, spit out a particle and turn into something else, you can never say when it is going to do that. You can say there is a probability in the next half-hour or twenty minutes, but it might be the next second. And probability comes into everything in quantum physics and so Einstein came up with that famous comment 'I cannot believe that God plays dice with the universe', but all the evidence is that God does play dice. The universe does work on probabilistic principles. It was something he could not come to terms with. It is an example, as with all scientists, that they get old and set in their ways and there comes a point when they cannot accept new ideas.

Roger Penrose takes up the point.

Well, it is a fact that when you make a

> measurement in quantum mechanics there is no theory that tells you what will happen. If you know the initial conditions, it just tells you certain probabilities of different results and Einstein thought that could not be the way the world was designed.

I asked him whether Einstein fought against it for reasons to do with his view of science or for other reasons, for instance to do with his character or what one might call, if not religion, his sense of the world.

> I think sense of the world, yes. That is one of the very striking things about Einstein. He had this tremendous insight into the way the world works and this insight in his younger years was proved to be very profound and accurate.

Many people believe Roger Penrose has made the greatest contribution to general relativity after Einstein, because of his work on the theory of 'black holes': he came up with the notion of 'event horizons' the regions within which light cannot escape and, with Hawking, proposed there was a space–time 'singularity', a point having mass but no dimensions at the centre of a black hole. Professor Lee Smolin, for instance, the young American physicist, argues that Penrose is the most important scientist since Einstein. I asked Roger Penrose – what next? Physicists seem to be predicting enormous changes in the way we understand the universe in the very near future.

> When things happen is always unpredictable, but I think there is hiding in the wings somewhere a major revolution, which involves the union of general relativity or the combination of general relativity with the rules of quantum mechanics. People usually refer to this as quantum gravity and there is a great deal of activity in the subject of quantum gravity. My own view is that that is not quite the right way of looking at it, that the rules of quantum mechanics must also be changed. So, in a way, I think what I am saying is rather in sympathy with what Einstein would have liked. That is to say, quantum theory is not the ultimate theory with regard to physics, which is, I believe, what many people think. Everything, in their view, has to accord with the rules of quantum mechanics. I say, no. General relativity will not completely accord with the rules of quantum mechanics. The rules of quantum mechanics will have to bend, as will the rules of general relativity. So we want a much more democratic union between these two theories where they will both bend. I think that once that theory comes, it will have major implications not just in the places people think, such as in the Big Bang or collapse, but I think that the fact that the rules of quantum mechanics need modification will begin to

> have big significance. I think there will be a major revolution coming.

To go back to one of the lesser threads in this book, I asked Roger Penrose two things: first, whether he anticipated this coming through an individual, as major analyses came through Newton and then through Einstein, and second; what will this revolution lead to?

> It is a good question. Again it is hard to predict the future, but I think it is the kind of thing that probably will come through an individual. We have in the early physics of this century two examples of physical revolutions. One is Einstein's general relativity, major advances, the other is quantum mechanics. Now, general relativity was basically the result of one man's thinking, whereas quantum mechanics was an enormous team effort, and I think what you are asking me is: do I expect this revolution to be more like Einstein's general relativity or more like what happened with quantum mechanics? I expect it to be more like Einstein's general relativity mainly because what I do see is it requires a completely different way of looking at quantum mechanics, in the same sense as Einstein's general relativity was a completely different way of looking at Newton's gravitational theory.

Einstein died in 1955 of heart failure, shortly after calling for his pen and latest page of calculations, according to his friend Abraham Pais. But his story is not yet over. Scientists today are still searching in the same way that Einstein did for answers to the questions that he first posed. Paul Davies comments on this.

> I might say as a philosophical point that people often complain about the scientists. They say 'You scientists, you think you know everything, you come up with this theory and then in a few years later that is overthrown, and you say something different. Newton thought he had it all wrapped up and then Einstein came along and showed that was all wrong. Someone will come along and show Einstein is all wrong. We cannot rely on anything you say.' Well, it is not a matter of being right and wrong.
>
> I do not think science is about truth, I think it is about describing the world in a reliable manner. Newton's mechanics, Newton's laws of motion and his notion of space and time is good for most purposes. Einstein's is a better description of the world which incorporates Newton's description. It is not that Newton was wrong, it is just that his theory is limited in its scope. It is inadequate in scope, mainly at the speed of light. I am sure that in the fullness of time we will find another theory which will incorporate Einstein's theory of relativity. That will not make Einstein wrong, it

will just mean that we will have a better theory.
I like to believe that out there in some abstract
mathematical realm there is the correct theory,
the exact theory that would fit nature like a glove
and that we are moving towards that in successive
approximations. And who knows? One day we may
get there.

Francis Crick

(1916–)

and

James Watson

(1928–)

1916 Crick born in Northampton, England.

1928 Watson born in Chicago, United States.

1932 Crick enters University College, London to study physics.

1940 Crick marries Ruth Dodd. They have one son.

1940–5 Crick works for the Admiralty on naval mines.

1943 Watson enters University of Chicago at the age of fifteen to study zoology. Graduates four years later.

1947 Crick divorces Ruth Dodd.

1949 Crick joins the Cavendish Laboratory, Cambridge, to work on X-ray crystal defraction. Marries Odile Speed. They have two daughters.

1950 Watson receives PhD from the University of Indiana and awarded Merck Fellowship to work in Copenhagen.

1951 Watson joins the Cavendish Laboratory, and meets Crick.

1953 Crick and Watson publish four papers on the structure and function of DNA.

1955 Watson goes to Harvard as Assistant Professor of Biology.

1961 Watson becomes Professor of Molecular Biology at Harvard.

1962 Crick and Watson awarded the Nobel Prize for Physiology or Medicine with Maurice Wilkins.

1968 Watson marries Elizabeth Lewis. They have two sons.

1976 Watson becomes Director of Cold Spring Harbor Laboratory, USA.

1977 Crick becomes Professor at Salk Institute in California.

1988 Watson becomes Director of the Human Genome Project, resigning in 1993.

The Meaning of Life?

As a teenager at the time of James Watson's and Francis Crick's world-shaking discovery of the molecular structure of DNA in 1953, I remember being intrigued by reports that we had discovered the secret of life. Yet on the day of what has been described by the scientist and writer Peter Medawar as the greatest scientific event of the twentieth century, James Watson underplayed its significance in a letter he wrote to Max Delbrück, a fellow scientist:

> *In the next day or so, Crick and I shall send a note to* Nature *proposing our structure as a possible model. If by chance it is right then I suspect we will be making a slight dent into the manner in which DNA can reproduce itself. I prefer this type of model over Pauling's, which, if true, tells us next to nothing about the essence of DNA reproduction. We would prefer your not mentioning this letter to Pauling. When our letter to* Nature *is completed we shall send him a copy.*

Peter Medawar wrote that *'It is impossible to argue with someone so stupid as not to realise that Crick and Watson's discovery was the greatest of the century*'. Others have been only a little less enthusiastic. Richard Dawkins, for example, told me that he thought it 'a good candidate for the greatest discovery of the century'. No one doubts its importance. Crick and Watson achieved a revolution in biology with their discovery of the double helical structure of DNA (deoxyribonucleic acid), providing the broad answer to the question of how genes replicate and carry information, and effectively beginning the whole new science of 'molecular biology'.

Of all the giants in this series, James Watson is the only one whom I interviewed. In manner and temperament I found him an unusual man – talking in sudden starts, abstracted for some of the time, yet always happy to try to explain. I began by referring to his book *The Double Helix*, in which he wrote '*Science seldom proceeds in the straightforward logical manner imagined by outsiders. Instead its steps forward (and backward) are often very human events in which personalities and cultural traditions play major roles*'. His description is interestingly similar to Poincaré's. How, I asked him, did it apply to the story of his and Crick's identification of DNA's structure?

> It was a story of four people, Maurice Wilkins, Rosalind Franklin, Francis Crick and myself. Each of us had quite distinct personalities which affected the final answer.
>
> Francis Crick was an extraordinary person. I would guess you would say he was a product of the

middle classes. His father was in leather goods. He had a factory in India and a factory in the Midlands. Francis was extraordinary, the way he could speak and his bravado and everything. He seemed like a character out of Shaw, the way he dressed, his style. I compared him, initially, to Henry Higgins in *Pygmalion*.

I was aided by the fact that I was not at all inhibited in going around asking what people were doing. Francis would probably hesitate to go to King's College, London, where Maurice Wilkins and Rosalind Franklin were working, too often because he would think they wanted to know what *he* thought. And *they* would think *he* wanted to know what *they* were doing. But as an American it just seemed natural to show up without appointments and letters, just to pop in. I was not constrained by good manners. It was not that I thought I had bad manners, but as a child I remember thinking manners are just something that keep you from really living, saying what you think, getting things done, saying something is good when it is not good.

Francis Crick was born in the middle of the First World War into a middle-class family near Northampton. His father ran a shoe factory. He showed an early interest in science – doing experiments at home when he was about ten – and read physics at University College, London, gaining a second class degree. Having worked for the

Admiralty during the Second World War, he decided to study biology and got a job in the Strangeways Research Laboratory at Cambridge University. Then in 1949 he had, according to his autobiography, a stroke of luck – he was offered a post in the university's Medical Research Council Unit at the Cavendish Laboratory, which was headed by Sir Lawrence Bragg.

James Watson, who was born in 1928, entered Chicago university at the age of fifteen to study zoology and then studied bacterial viruses for his PhD at Indiana University. Early on he wanted to find out what a gene was, which took him to Copenhagen to study bacterial metabolism. He moved on to Cambridge in 1951, joining Francis Crick in the team led by Max Perutz despite initially having no grant. In 1953, at the time of the breakthrough, he was still only twenty-five.

The science of genetics was barely a century old in the 1950s. Gregor Mendel, the botanist, experimented with crossing varieties of garden peas in the middle of the last century. He concluded that his plants inherited two hereditary components, what we now know to be genes, one from each parent. It was not until the early years of this century, when Mendel's theories were developed and combined with those of the great evolutionist Darwin, that their full significance was grasped.

Linus Pauling (1901–1994) is also a very important figure in the history of DNA – some would say he was the outstanding chemist of the twentieth century, and one of his massive contributions came from being the first to use X-rays to work out the shape of proteins. His work on the structure of 'biomolecules' and on hydrogen bonding formed the basis of Crick and Watson's world of DNA.

By the 1950s the race to discover how this hereditary information was passed on was well under way, with the teams at the Cavendish Laboratory in Cambridge and King's College in London the front runners. They both suspected that our genes were made of DNA, the chemical substance found in the nucleus of every cell of every living organism. The person who had worked hardest to collect the crucial data, using the latest X-ray techniques, was neither Crick nor Watson, but Rosalind Franklin of the King's College team.

Crick and Watson did not see eye-to-eye with the independently minded Franklin. Watson admitted at the time that he thought the best home for a feminist was in another person's laboratory. Someone who has since taken sharp exception to that is Evelyn Fox Keller, Professor of History and Philosophy of Science in the programme Science, Technology and Society at the Massachusetts Institute of Technology.

> Rosalind Franklin was a kind of heroic scientist. She really believed in doing the work for its own sake. She just loved science. And she was an extraordinarily good scientist. It is a tragedy that she lost out in the official history of the subject. It is particularly unfortunate that the image of Rosalind Franklin that is passed on to the

> generations is so very much formed by Watson's *The Double Helix*. It is a terrible misrepresentation, as I think Watson himself would admit today. Nobody called Rosalind Franklin 'Rosie', for example. She was a woman of extraordinary integrity and dignity. Obviously, there will be continued controversy as to how quickly she would have arrived at the same formulation that Watson and Crick did. But I am not sure that that is the point. I think that a more interesting point is the extraordinary focused attention of Watson and Crick on DNA. As Watson is quoted as having said in that wonderful film *The Race to the Double Helix* – 'DNA is gold' and he was going to go for it.

James Watson has, of course, seen the film himself and has strong opinions about why it was Crick and he who discovered the structure of DNA, not Rosalind Franklin.

> I thought at the ending of the BBC movie, which really focuses on Rosalind, she simply comes out as the more appealing of the two of us because she accepted defeat in such a strong fashion. She did not whine. Rosalind was not a whiner, which certainly made life easier for Crick and me because there were a lot of people who thought we stole the problem from her. Rosalind realised that she just did not pick it up and run with it.
>
> There were three reasons why Rosalind did

> not get the answer, and if any one of the three had been different she would have got the answer. One is that she did not live DNA, in fact she was prepared to stop working on it. She should have stuck with DNA. She probably did not think DNA all the time, the way I did. The second reason was that there was no one who would protect her, she had no patron. Normally, you know, if things go bad and you are knocked down, you go home; but she could not go home because she had essentially burned her bridges with her father. And the third was, she really could not collaborate easily with outsiders. She had said that DNA was not a helix because of some slight asymmetry in the X-ray pattern and Francis had said this was a red herring, and it was silly, it was just how the crystal was formed, it had nothing to do with the underlying symmetry of the molecule. And she was annoyed that Francis had said that she was being silly. But she should really have talked to him because, as I discovered, it is hard to be successful in science unless you talk to your opponents. You have got to know the other people, even though you may find them objectionable in every way. You had better know what their arguments against you are.

I asked James Watson if he thought Rosalind Franklin had suffered from being a woman in what seems from his book to have been an almost exclusively male world.

> I think she chiefly suffered from an unbending personality which could not make friends easily. There were something like fourteen women working at King's College at that time and she did not become friends with any of them. She was a solitary person and she did not really talk to anyone.

He explained how Maurice Wilkins and Rosalind Franklin began working together on DNA.

> Wilkins had read Schrödinger's book *What Is Life?* and, as a physicist interested in biology, he knew if you could get the gene you were probably going to do something fundamental. So Wilkins began to focus on DNA using various physical techniques to say something about its structure. He had got a sample of DNA from the Swiss biochemist Siegner, which he exposed to X-rays and got what we call 'crystalline photographs'. It was a big step forward. But he was not a trained crystallographer and he had done it sort of as a lark.
>
> And when he was in the States, Rosalind Franklin was hired. She was a chemist who had gone to Cambridge and then had been to Paris for four years, and then I think she thought she really had to come back to England and had got a job at King's, but not to work on DNA. Sir John Randall – her boss – thought that really she was better trained

> than Wilkins to carry out the X-ray work on DNA. He essentially told her 'You should work on DNA', although Wilkins had started it. So Wilkins got home and discovered he had got a collaborator, but not of his choosing, and very soon they got on each other's nerves, which I think is an understatement.

Richard Dawkins, Oxford Professor of the Public Understanding of Science, disapproves of the way Rosalind Franklin was treated.

> In some ways I would have liked to have been alongside such exciting work. But if you read *The Double Helix*, there are all sorts of disagreeable aspects to it. I suppose the worst of it is the way the scientific society treated Rosalind Franklin. I do not mean the way Watson and Crick treated her, I mean the way she was not even allowed to go into the Common Room in her own institution and therefore could not talk shop with colleagues. I would really, really not like to be in that atmosphere. I am extremely glad we have left that behind us. But setting that aside, it would have been very exciting.

Clearly Watson and Crick must have got on, or they could not have worked together so closely. I asked James Watson what had made them such a good team intellectually and why it was so important to them both to be at the Cavendish Laboratory.

> The Cavendish was as good a place in the world as there was to do X-ray diffraction. Essentially the field had been opened up by Lawrence Bragg and his father. The first structures were solved by the Braggs. So it had a collection of people who did X-ray crystallography. That annoyed the physics group who thought that Bragg was not continuing the tradition of nuclear physics at the Cavendish, the greatest experimental laboratory that ever existed in the world. It is still true. Bragg was in fact moving partially to his own interest. Bragg, I think, appreciated that the real challenge was to solve bigger and bigger molecules which really were the proteins.
>
> I brought to our partnership a sort of gossip about the world of bacterial viruses and the people who were using viruses to try and understand what the gene was. Francis Crick knew he should be interested in DNA. He was working on protein but if the gene was DNA, then the gene was more important than anything else. Francis Crick wanted to think about important things. In talking to me he could talk about something which gave him a diversion from his own thesis on haemoglobin which was not making any big advance.
>
> We liked to think the same way. I did not

> have the mathematical knowledge to think the way Francis did, but we both were bored by dull things and we were under the impression that most people did dull science. You know, why should you not do something that was fundamental? We both had the courage to say 'There's a bright chap doing something dull'. Lawrence Bragg said do not rock the boat, or ship, or something, but we were both boat-rockers. If we thought we were in a mess or we thought that collective opinion was wrong, we never went along for consensus. We were not consensus thinkers.

Crick and Watson gossiped about their work constantly, to each other and to anyone who would listen. Unlike Rosalind Franklin, they preferred analysing other people's results to actually experimenting and gathering data themselves. At one point they exhibited a possible model of the structure of DNA based on information that Watson had inaccurately picked up from one of Rosalind Franklin's lectures. When she saw the mistake, Franklin coolly pointed out their proposed model contained ten times too little water. Humbled but undaunted, the duo searched on until, one Saturday morning in the spring of 1953, Watson stumbled on a simple structure, looking something like a twisted rope ladder, that seemed to work. He recalled the moment of discovery when I talked to him. I asked if they had rushed off to celebrate.

> I think we were pretty excited. You don't want to

> wake up before the dream vanishes, it's a feeling. And then . . . I'm sure we told people at lunch in The Eagle, because we always had Saturday lunch at The Eagle. That was the pub which was about three hundred feet from the doors of the Cavendish, so pretty close. And then I was supposed to go to Paris and in Paris I learned of some new evidence that made it even more likely. Then we wrote up the manuscript.

The pair proposed that the DNA molecule was composed of two simple spirally wound or helical chains. The DNA chain nestling in *each* microscopic human cell is a staggering two metres long when stretched out. If all the DNA in a single human being were unravelled, it would extend from the earth to the sun and back again. Francis Crick explained it on the day in 1962 when he, James Watson and Maurice Wilkins were awarded the Nobel Prize.

> DNA is a polymer. That is to say it has a regular, repeating backbone with side groups called 'bases' projecting at regular intervals. However, all the bases are not the same, there are four kinds of them and the genetic information is conveyed by the precise order of the different sorts of bases along the DNA. In other words, the genetic message is written in a language of four letters. Incidentally, the total length of the message for man is not short:

> it is probably more than a thousand million letters long.

There is no doubt that Crick and Watson's discovery has revolutionised biology over the last forty-five years. In less than half a century, medicine has made great strides forward as scientists have recognised that the causes of many illnesses are genetic in origin. There have been benefits in other spheres too, from palaeontology – where it enables deductions about the physical appearance and way of life of, for example, Neanderthals to be made from bones – to crime-busting.

But Evelyn Fox Keller thinks we should be careful not to overestimate the implication of Crick and Watson's discovery. Too often people have claimed that the understanding of the structure of the gene can be regarded as a theory of everything, the solution to all our biological problems, whereas she considers that DNA is really just rows and rows of data.

> The identification of the structure of DNA is, let me say, *the* most important biological milestone of the twentieth century. The identification of the structure of DNA gave tremendously powerful support to the belief that had been gaining hold for half a century, the belief that there was a thing called the gene which would explain both heredity and development. The DNA was the master molecule, it was the executive suite of the cell of

the organism. And from that point of view, it was inevitable that people would want to get at that executive office, get at that master molecule, learn its secrets, learn everything about it. But the story was always more complicated and there were always people who were trying to say this.

The success of Watson and Crick was so dramatic that it was not possible to hear an alternative story. Now it has become more possible because what we know has become more complicated. We know so much more. Now, when Watson said 'There's gold in DNA' he was right, but one might ask 'Is the understanding that we are going to get from DNA about biological processes, particularly about biological development, unlimited?' I think what we have learnt is that there are real limits to what DNA is going to tell us.

Richard Dawkins has another view.

There is a tendency, which I do not altogether approve of, to say that if something is simple and tidy and clear-cut there must be something wrong with it. There are people who want the world to be a complicated messy place and so, when a Watson or a Crick comes along and shows that what had been messy is beautifully and elegantly and supremely simple, there is a sort of resentment.

John Maynard Smith, Emeritus Professor of Biology at the University of Sussex, does not hesitate to praise the achievement of Crick and Watson.

> To discover the chemical basis of heredity is profoundly important when you bear in mind that the whole not only of evolution but the whole of development, the whole of biology, depends on that property of heredity. To say that you now actually know how it works, is profoundly important.

Although Mendel had discovered that genes did not mix in the offspring as Darwin had suspected they might, his work was paid little attention at the time and the prevailing idea in the middle of the nineteenth century was that, for example, a short woman and a tall man would have a child of medium height. That theory was later discounted in favour of the child randomly inheriting the gene from one of the parents. What Crick and Watson had done was to show how this genetic information was passed on through the DNA in our genes. Richard Dawkins explains this further.

> We already knew because of Mendel that genetics was digital and not analogue. That is to say, genes do not blend, they are either there or they are not, it is like beads on a string. What Crick and Watson did was to show that even within a gene it is digital. Even down to the finest minute structure

of a gene, everything is digital code. It is just like computer language. The only significant difference is that it is not binary but it is quaternary and that is a trivial difference. So, whereas before Crick and Watson came along it was still possible for people to wax a bit mystical about genes and to say there is something deeply mysterious, a sort of life force about them, that turns out to be total nonsense.

The gene is just like a piece of human computer tape. You can transcribe a gene, it is being done at this moment for the Human Genome Project, into print on to computer tape. You could get the whole human genome and put it into a library in the form of ink on paper and then at some future date, some future century, take down the books from the library, transcribe that ink on paper information back into DNA, and in principle you could regrow an identical twin of the individual whose data you originally used.

So that is a supreme revolution in our view of life and it has clearly been the basis for an immense amount of work in medicine, in biochemistry. The whole of biology has been completely turned upside down by Watson and Crick.

Crick and Watson also explained why early evolutionists were wrong in thinking that characteristics acquired during a lifetime could be passed on to the offspring. For example, a naturally pale-skinned person, who spends their life in the sun, will not pass on tanned skin to their children. John Maynard Smith develops this point.

> One of the things that comes from the Watson/Crick discovery and the work that immediately followed it, is that we now understand why acquired characters are not inherited. Now this is a bit odd. We are very used to information being transmitted by machinery. We are used to telephones and record players and stuff which is transmitting information in the way that the genes do, and some of the machines we use are reversible, like a tape recorder. On the other hand I can shout my head off at a record player and it is not going to cut a groove in the record. It just turns out that the machinery whereby genes influence development is like a record player. It will not work backwards. You can change the organism and it does not enable it to change its genes.

A mere forty-five years after Watson and Crick's breakthrough, we are well on the way to understanding what each of the hundred thousand genes in our bodies is for, thanks to the Human Genome Project directed until 1993 by James Watson. It is a vastly ambitious venture and one which may enable us eventually to screen out genetic diseases, such as cystic fibrosis. But it has caused immense controversy. Are genetically engineered human beings next on the agenda? Evelyn Fox Keller is characteristically cautious.

> We do have to get away from the idea, the expectation, that we are going to hold the future in our hands so simply. They can do a lot of genetic manipulation but the impact of that manipulation, I think, so far has been far more important to the bio-tech companies than it has been to the population as a whole.

John Maynard Smith takes a wider view.

> I am wondering whether the objection to the Human Genome Project does not arise from what is quite a healthy fear of fiddling. So I think, in this case, it is not the knowledge that people are frightened of but what use that knowledge might be put to and I think people are right to want to be involved in the decisions.

That, I thought, brought us back to Evelyn Fox Keller, who says that even when we know all about human genes, there is something more. John Maynard Smith agreed.

> She is absolutely right about that, absolutely right. Even if we did have the complete sequence of the human genome in front of us, it would not answer most of the questions that I want answered about biology. It would not explain to us how you and I were able to develop from an egg. It would not

> explain how, having developed, we were able to learn to talk, to run about and so on. None of those things would emerge. All the Human Genome Project will do for us, is a bit like providing us with a dictionary that we can look things up in. Obviously, if you are trying to learn a new language having a dictionary helps, but it does not explain the meaning of the novel you are reading.

In that sense a lot of people are over-estimating its anticipated value, I suggested to John Maynard Smith, because a lot of people think that once we get this map of the body we can take a stab at it with a pin and say 'I do not want that bit, or this. I want to be a different person.' We can be our own Frankenstein, our own monsters; that is what some people fear and others hope.

> I think if they believe that then they are mistaken. I do think it would help us to identify certain specific things that we would prefer our children not to have, and possibly alter them. But to say that I would like to have my child to run a hundred yards quicker than anyone else or to sing beautifully or to do mathematics – it is a lot of rubbish.

One of the most controversial developments to have emerged recently as a direct result of the Crick and Watson discovery is the cloning of Dolly the sheep. I asked James Watson if he ever dreamt that his work would have led to this and what he thought about the very real possibility of being able to clone human beings in the future. Could we do it?

> I guess reluctantly I would say, yes, but I would hope that it would be otherwise. It would change the way people reproduce. You could say I am not going to take the chance of just the joint union of a sperm cell from my father and an egg from my mother, because we do not know what is going to come out. We will just take a cell from a body we know that has been pretty successful. Many people have children who get the wrong combination of genes and that is very tough on the child, it is very tough on the parents. I think it is irresponsible to let a child be born with just suffering ahead in their life.

Whilst John Maynard Smith and Richard Dawkins broadly agree that there are important medical benefits, they do not see eye to eye on the ethical problems raised by the prospect of cloning human beings. John Maynard Smith is entirely against it.

> Think of it from the point of view of the cloned human being. I mean, would it not be absolutely awful to be a little cloned Pavarotti and your parents would be standing saying 'When, you little beast, when are you going to sing nicely?' I mean, the expectations would be impossible.

Richard Dawkins disagrees.

> There is a tendency to a knee-jerk reaction to anything new, particularly, it seems to me, where it concerns human reproduction, which seems to be a very sensitive issue. There are people who will automatically say 'Cloning is wrong, human cloning is wrong, it has got to be stopped'. I am very prepared to be persuaded that it is wrong and ought to be stopped, but I have not actually heard any good arguments in favour of that proposition.

But would any benefits come from it, I wondered?

> Benefits? You have to ask: benefits to whom, I suppose. I mean, speaking purely personally and for sort of curiosity reasons, I would love to be cloned myself. But as to benefits to humanity as a whole, I find it hard to actually think of benefits of cloning a present adult unless you count deciding that certain individuals, say Einstein or Nelson Mandela, are such magnificent individuals that it is worth having another of them.

The question is whether such a clone really would produce another Einstein or Mandela.

> You would get a jolly good start towards another of them. You certainly could not guarantee to get

> another of them. It would be like identical twins and identical twins, very often, turn out to have similar talents but not always, and people raise all sorts of other things, like 'Well, how do you decide who are the virtuous individuals who deserve to be cloned?' The spectre of Saddam Hussein producing regiments of little Saddams is raised. So that, yes, these are issues that must be discussed.

I asked James Watson whether he thought that he and Crick were standing on any giants' shoulders.

> I guess we were standing on the shoulders of Lawrence Bragg and Linus Pauling because they were probably the two dominant people. Pauling to understand the nature of the chemical bond which Bragg was not all that interested in. Bragg was not a chemist, but Bragg's Law was the basis of understanding in solving the molecular structure. So we were standing on the shoulders of Bragg and then Pauling.

Since their now world-famous discovery, Watson has stayed focused on molecular biology while Crick later left the field, after further important work, in favour of neuroscience, the study of the brain. In a rare get-together, twenty-five years ago, Watson explained that, in writing *The Double Helix*, he wanted to show that science is an everyday operation performed not by geniuses but by ordinary mortals and punctuated more often by long periods of inactivity than by moments of earth-shattering inspiration. I think it is worth reproducing their conversation at some length: the contrasts and insights are fascinating.

> WATSON: I get bored by people referring to me as a genius, or something like that. I know what I am, so I just wanted to put the whole thing in perspective of what we thought about or didn't.
> CRICK: But when you put it in perspective, Jim, you did make it a little bit easier than it was. What you were trying to do at the time was to make out we just were not cold characters in white coats and we were human and that you did, I think, admirably. But, owing to the fact you wanted to make it readable to people who did not understand the technical side, a lot of the technical bits were left out, so it sounded as if, you know, anybody could have done it.
> WATSON: Well, if you want to take the eighteen-month interval, between when I arrived in Cambridge and when we got the structure, and you ask what percentage of our actual working days were spent thinking on DNA during that eighteen-month interval, I do not think you could come up with more than three months' worth, probably less.
> CRICK: I absolutely agree, but I do not think that is what matters. I mean, the moment of conception is often brief.

WATSON: But most of the time we were doing something else, there were enormous periods of just drinking coffee or taking walks, wondering why we could not think of the right answer.
CRICK: But even people who do experiments, most of their experiments are no good and they do not work and they have to do them over again and so forth. It depends what you call work, you know.
WATSON: There were just long intervals in which we were stuck, in which the important thing was that you had enough sense to stop thinking about it, so that you did not get total frustration, and thought about something else. And so there were long intervals when we could not do anything, we suddenly came back and then roughly in six weeks the whole thing went through very fast, but before that . . .
CRICK: That is true.
WATSON: . . . there were very long periods when you and I were thinking different things and that was one of the things I wanted to put across.

International celebrities they may be, but with their discoveries still in living memory, Crick and Watson have not yet had time to become the remote textbook giants like Archimedes or Newton or the other figures we have examined in this series. I wanted to end our interview by asking James Watson what he thought had made him a Nobel Prize-winning scientist. Why do scientists do science?

> I just like to know why things happen and I think that is probably something we have inherited. Curiosity about things, why things happen, can prepare you for how you live in the world. It has great survival value, this sort of curiosity and it is a question of how your curiosity is directed. Many people are very curious about things, are obsessed about things, which you could say have no consequence.

An answer which, in my opinion, would have been given by most of the scientists in this book.

Where Are We Now?

Where Are We Now?

THE QUESTION of where we are now epitomises the need of the non-scientist to get to the nub of the matter without all the hard work. Any answer which would satisfy a scientist would, I am sure, take into account innumerable instances and imponderables and be much more concerned with minutiae than the broad sweep.

Yet, as we have seen, today's scientists are not unwilling to risk generalisations understandable by the layman. It was their willingness to take risks and their generosity to share their knowledge which drew me into their pursuits in the first place – although I would make no claim greater than that I have stood on the sidelines and paid attention. What I have come away with is a map – primitive, no doubt – a feeling for certain centres of intellectual excitement but, above all, a sense that science is not the cold outsider with all the answers but a shifting, fallible, human exercise intent on examining the meaning and the purpose as much as the structure of life today.

I asked several of those I had interviewed to give me their snapshots of the present and the future. Jocelyn Bell Burnell is convinced that all the physics she has ever studied covers only five per cent of the universe. The other ninety-five per cent is not only undiscovered, it is 'totally different and it is a fresh start, a big fresh start'. Sir Martin Rees spoke of 'a frontier that will exist for ever, in my opinion, which is to understand the complexities evolved in our universe over its fifteen billion years of history'. Paul Davies feels that he will not in his lifetime quite 'get to the final goal. 'Humanity,' he claims, 'is coming close to glimpsing the mind of God, if I could wax lyrical, but that will have to be left to a different generation'. John Maynard Smith brings a cold splash to the subject. 'Trying to predict the future is a mug's game,' he says. Despite that douche, this is what this final chapter will partly attempt to do.

What it does not do is try to answer what might be the most pondered question about science – namely, will it be a force for good or for evil? Currently the emphasis seem to fall on science's monstrous possibilities – simplified and sensationalised, they are the meat of the most widely popular films of the day and the stuff of much fiction, some of it reaching a colossal readership. Gothic horrors, ghost horrors, even the horrors of war present no competition to the 'horrors' of unchecked science, of unknown creatures from a future 'scientific' world, and of unelected scientists fanatically shaping the future of the world. New cures for diseases, advances in food production and all the technological conveniences of transport, communication, domestic ease and public entertainment are acknowledged and enjoyed and unceasingly desired, but somehow they live in the shadow of that view of science which sees the dark as the true landscape of this enterprise. Yet what this book has unfolded, I think, is a story of enhancing the lightness of being. Discoveries have been made in that good faith which comes from the determined pursuit of truth and, if they have later been abused, then that is not the fault of the messenger. For in one sense that is what scientists are – messengers from us to and from the unknown. *Homo sapiens* will never stop seeking to find out what he does not know. However, how he uses that knowledge is an allied but in the end a separate concern from this history.

Our subjects stand, as Sir Isaac Newton put it, on the shoulders of the giants before them: Archimedes, the inspired Greek mathematician; Galileo, the Columbus of the stars; Newton, the difficult genius who invented the scientific method we use now; Darwin, the seemingly ordinary young man who came up with such an extraordinary theory that it continues to inspire scientists in many fields today. The story of their success in discoveries suggests that science can offer some of the last great certainties: difficult questions remain, of course, but science can at least offer answers, solutions and progress. Where once most people looked to faith for comfort, today many optimistically expect the next generation of scientists to add to the knowledge we already have and improve on it. They have faith that the future will only be better. That is one view.

This brief look at the history of science has taught me, if nothing else, that nothing is so straightforward. No brilliant insight has gone unchallenged and the scientific and the ethical do not run on parallel lines. The debate on the future of science is intense. Lines are drawn. Arguments and armour are in place.

John Horgan is a writer on the journal *Scientific American* with a strong belief that science as we know it is drawing to a close.

> Science, right now, is a victim of its own success. It has achieved so much already that I think it is going to be very hard, increasingly difficult, to top what it has already done. We have already created this map of the entire universe from quarks and electrons all the way out to galaxies and quasars on the very edge of the universe. Physicists have shown that all matter consists of a handful of particles ruled by a few basic forces and they have created these very powerful theories of these particles and forces.
>
> When you look at the biological realm, we have Darwin's theory of evolution and modern genetics, which again create this very powerful framework and narrative for understanding all of biology. So what is happening right now, and it will continue to happen in the future, is that scientists are really just filling in the details of this map of reality and this narrative of creation that they have already built with their current theories. We are not going to have any of these gigantic revelations into nature that we had previously with quantum mechanics and

> the theory of evolution, general relativity or the Big Bang theory. Yes, science will continue, but its glory days are over.

Sir John Maddox was the editor of the scientific journal *Nature* for over twenty years. He is wholly against the Horgan camp.

> I do not agree at all. It seems to me that one can equally well argue that science is just beginning. The problems that lie ahead, many of which can be talked about and defined, are huge problems, as huge as any that have been tackled in the past century. And more than that, of course, I believe that what constitutes progress in science is not so much the next gee-whizz discovery, it is learning how to ask questions that even Aristotle asked, more perceptively, more incisively and more meaningfully. I think that the constant deepening of the level at which we are now asking questions of nature is one of the remarkable things about the progress of science.

I asked John Maddox if he thought that the same questions would always be asked.

> Let us take a particular case. How does the brain work? Aristotle tackled that. He thought that mind resided not exclusively in the brain but in the blood

and in the heart and so on. Now the question of 'Where is the mind and what is it?' is still a taxing question which all the neurophysiology of the past century, marvellous work though it is, has not answered, and it seems to me that that question is going to keep on being asked for centuries until, at some point, we will have an answer that satisfies everybody. And that will be the end, of course. Because by then people will have got on to asking questions that we ourselves are not smart enough to formulate.

Are there no final answers because there are no final questions?

I think there are no final questions. I think the issue of whether there are final answers is an interesting philosophical question. It seems to me that we have now learned that science is a process of successive approximations to the truth. Whether the truth is absolute or not remains an open question, it seems to me. And on questions like 'How does the mind get into the brain?' or on questions like 'How does the genetic composition of a person evolve in the course of time?' – these are questions that may not have cut and dried answers in the old-fashioned sense.

I wondered whether he thought the feeling that some people

have at the end of this century, that the fundamental map has been drawn, was matched at the end of last century?

I think that is very interesting. A century ago was one of the great triumphalist periods in science; that was when, at the end of the nineteenth century, people thought 'It is all done now. Newton's mechanics has stood up fine, we have got better mathematics to go with it, which makes it possible to solve pretty well any problem you care to formulate'. People thought that Darwin had explained the huge diversity and interest of the living world, and so it seemed that science was set for contentment.

But look what happened. Within five years the subject of physics had been turned upside down, with the discovery of quantum mechanics by Planck and Einstein; relativity by Einstein; a new theory of gravitation by Einstein, and then the quite marvellous foundation of quantum mechanics, between 1900 and 1925. And there never has been such a revolutionary period in science. One can see the beginnings of that revolution in the nineteenth century even while people were boasting about the splendour of their achievement. They knew that there were X-rays – a great challenge to classical physics – radioactivity and the electron, which seemed to be part of an atom and therefore suggested that atoms are not indivisible as everyone had assumed. And so the origins of the marvellous revolution that we had at the beginning of this century were to be found in the contentment of the nineteenth century. And it is my contention that exactly that kind of forewarning of revolution is now apparent in modern science.

For example, there is a warning – more than a warning, there is an indication – of a great need to change the way in which we regard cosmology. At present it is supposed that the universe began with a Big Bang something like ten or twenty thousand million years ago, and the Big Bang appeared in empty space, it was not an explosion of the ordinary kind. It created both the matter that is now in the universe and the space and the time that universe expands. That is the Big Bang story.

The Big Bang has been challenged from both the experimental side by the measurements of the ages of stars in our galaxy, which turn out to be greater than the age of the universe itself – that is an impossibility, of course. But, quite apart from the observations, there is also the enormous difficulty of trying to make the laws of physics work when the Big Bang was happening. It is a question of putting together Einstein's theory of gravitation with quantum mechanics. These two great pillars of physics created in the beginning of this century are irreconcilable when looking at how the Big Bang developed. So there is a tremendous need to

> make a marriage between gravitation and quantum mechanics.
>
> Take another field – the brain. How does it work? This again is crying out for a solution that nobody has yet provided. How did life on earth begin? We know when it began. Some people say it is not a problem in science because it was obviously a historical accident but, at least, it should be possible to demonstrate, and it is essential that it should be demonstrated that life can begin with the kinds of molecules which were on earth and that that life could then evolve into what we are. These are three problems untackled as yet.

Jocelyn Bell Burnell agreed with John Maddox that the revolution in nineteenth century physics is about to be repeated at the end of this century.

> I think it is fair to say that there has had to be quite a lot of consolidation since that revolution. I actually believe that there is another revolution imminent. This is getting on to a very big subject, but there is evidence, when we look at the universe around us, that there is a lot more gravity than we suspected in the universe, which implies that there is a lot more matter than we suspect in the universe. We do not know what this matter is and it is not visible, we cannot see it, but it must be there. Indeed, the stuff we know about probably

> only comprises about five per cent of the universe and the other ninety-five per cent is, in physicists' terms, something totally different.

John Horgan's response was rather like that of Mandy Rice Davies – 'Well, scientists would say that, wouldn't they?'. He believes, two and a half millennia after Archimedes, that we have become so caught up in the idea that science is always about progress that we cannot imagine anything else.

> There are very few people who agree with what I am saying. I think it is because we have had such astonishing progress in science, particularly in the last hundred years or even the last fifty years. We all were born into that period of amazing progress in pure science and technology that we have known nothing but that, and so we just assume that this is now an intrinsic feature of our culture.
>
> What I am saying is that if you stand back and look at it from an historical perspective, really it is much more reasonable to think that this period of explosive progress is an anomaly in human history. So that the only way you can believe what John Maddox believes is if you are a hopeless Romantic who just has faith in science as a religion and wants to continue to believe that it will continue for ever.
>
> General relativity imposes limits on how fast we can go through space, so forget about all those

spaceships in *Star Trek* or *Star Wars* going at warp speed, faster than light through the universe, and visiting other galaxies and other planetary systems – it is not going to happen if you accept what science tells us now. Evolutionary biology also keeps reminding us that we were not put on earth to discover great truths of nature, we are animals who turned out to have this peculiar capacity for figuring lots of things out. But there is no reason to think that we should continue discovering these profound new things for ever. There are certain limits that nature imposes on us just because it is so complex that we cannot comprehend it. Talking about the human mind, I think that is a really good example of that. For scientists like Roger Penrose, discovering the truth about nature is what makes life meaningful, and the idea that that process should stop at some point is deeply disturbing to them, even when their own theories imply that it is going to happen.

Let me just make it clear that I am not saying that we know everything that we want to know; that we have dispelled all mysteries from the universe. There is this great paradox that modern science poses, that the more we know, the more mysterious the universe becomes. For example, we have the Big Bang theory, which really is a very powerful and I think an absolutely true theory of creation. It tells us that the universe had some kind

> of creation event about fifteen billion years ago and is still expanding to this day. But where did the universe come from in the first place – where did the Big Bang come from? Where did the laws of nature, of physics, come from? There are really smart people who are trying to answer these kinds of questions, but their theories postulate phenomena that can never be experimentally validated. They talk about other universes that are out there in other dimensions, and little particles called superstrings, that exist not just in three dimensions but in ten dimensions – and all these wonderfully fantastic ideas. But if you press the scientists on whether they can ever have as much verification of these things as they have of the existence of electrons or galaxies or things like that, which we can observe directly, I think they will have to admit that they cannot, that these things will always be in the realm of speculation, so that makes them more like theology or philosophy than science.

The argument using the analogy with the last century is a potent one for many scientists, but John Horgan will have none of it.

> Nine times out of ten, when I present my argument that science is ending, the response I get, whether it is someone I meet at a cocktail party or a Nobel laureate in physics, is 'That is what they said at the

> end of the last century'. It is amazing the degree to which they dismissed this entire argument about whether there might be limits. There has been very little literature on this subject, just because of what I think is a pretty fatuous argument.
>
> First of all, physicists at the end of the last century did not think that everything had been wrapped up. If you talk to historians of science, they will tell you that scientists were debating the existence of atoms – you cannot get much more profound than that. But the big difference between now and then is just that we know more. We know that atoms exist, that there are electrons and quarks and things like that. We have got very good evidence that the universe is expanding. Since the end of the last century we have also heard what genes are made of – DNA – so the more you discover, the less there *is* to discover. I think the only way you can have an opposite point of view is if you think that science does not discover truths, it just goes from one point of view to the next. You have to have a kind of post-modern view of science to think that science is going to continue for ever.

One problem which may continue for ever is that of consciousness and beyond consciousness, I suggest, the imagination. We are nowhere near 'discovering' that in the sense that the structure of DNA has been discovered. Horgan has an answer even for this.

The human mind is the last great frontier for science and this is a common objection I get to my argument. How can we say that science is over when we are just beginning to study the human mind and we still have no basic understanding? My argument here is a bit different. In some of these areas I am saying we have really figured things out to a great degree and we are just going to be filling in details. When it comes to the human mind we have had at least a century of fairly rigorous scientific investigation since Freud.

Freud posed his theories of psychoanalysis at the end of the last century and developed them over the next few decades. We are still arguing over Freud's ideas now: in practically any book review section reviewers argue the pros and cons of Freudian theory. I think it is pretty clear that Freud is not very scientific. It is more like pseudo-science than real science. The fact that we are still arguing over his theories means to me that nothing better has come along to displace them.

I guess my feeling is that the human mind is just intractably complicated. People say such silly things when they are trying to explain consciousness that to me it says that this is something that is fundamentally mysterious. I do not think that science can really comprehend it in the way it has comprehended, say, heredity, which is basically an understood problem. So I think we are going to

> continue to get new theories proposed constantly until the end of time, but I do not think you are going to get that kind of convergence on a single theory that you have, say, in nuclear physics or genetics.

Roger Penrose is reluctant to see an end to science.

> I suppose the reasonable answer would be to say there will come a time, it is not in the foreseeable future, but one might say that ultimately sufficient understanding will be there. But I do not see any indication of that now. I think that it has always been the case that, when one thinks one has got close to the end, there are some huge things which are simply not appreciated at all. I do not see any end to it at the moment, but it is hard to say that there will never be such an end. I think that in some sense it is harder to say also that the world makes sense without there being some kind of an overall view of what it is all about, but I do not think we are close to that overall view at the moment.

If you ask the scientists, there appears to be a consensus about where the big gaps in science remain, and they seem to me to be the same big questions which fascinate the non-scientists too. Richard Dawkins agrees with John

Horgan that one of those gaps lies in our knowledge of the mind.

> The greatest riddle I can see facing biology is the nature of human consciousness and it may be more of a philosophical problem than a scientific one. It is something that is arousing great interest at the moment. I do not feel I really understand what the problem is, let alone what the solution to it is, but I think that would clearly have to be my number one.

I thought that if Richard Dawkins did not understand what the problem is then that put the rest of us in rather a difficult position. It was important to discover what difficulty he had with understanding the problem.

> You could program a computer – indeed you *can* program a computer – to be very good at chess, Grand Master standard chess; to solve mathematical problems; to converse about a limited domain in ordinary language; to solve logical problems. There are all sorts of things you can make a computer do which look conscious, which might even pass Turing's test. Remember? Alan Turing (the British mathematician) made this thought experiment, that if you could have a person sitting in one room and a computer sitting in another room and another

human sitting in another room, and the human, the subject, was conversing by teleprinter, and he did not know whether he was conversing with a computer or a human, he could not tell which was which, then that is Turing's demonstration that it is conscious.

I can never decide, this is my dilemma, whether if I did that and I was finally told 'Yes, it was the computer', I would really believe that it was conscious. Because I could imagine a computer being programmed to mimic human behaviour in all its complexity and apparent intelligence, real intelligence, with great fidelity. I still could not quite bring myself to belief that it was having conscious experiences in the same way that a human being does. On the other hand, perhaps if it happened to me I would convince myself. In any case that is a tiny taste of why I think it is a difficult problem. I am not even convinced that I would know consciousness in a machine if I met it, which means one would have to generalise and say 'I do not really know that you are conscious, I only know that I am', and the only way out of that sort of solipsism is to say 'Well, we are very similar, we have come into the world by the same sort of process, we have the same provenance, it would be very, very strange if one of us was conscious and the other one was not'. And so, therefore, I believe that you are and I believe that all other humans are

> and I believe that chimps are, but I probably do not believe that worms are – well, why not? And do I believe that chess-playing computers are? Actually no, but at some point in the future they may be. These are all more or less conjectures and I cannot see an easy way to answer any of them with certainty.

Igor Aleksander, Professor of Neural Systems Engineering and Head of the Department of Electrical Engineering at Imperial College, London, is one man brave enough to tackle the question which frightens the non-scientists and irritates the scientists alike. He believes that the neurons in our brain are the source of consciousness, and he has developed a computer, Magnus, which he controversially argues allows us to see how those neurons might work and what, therefore, consciousness might be.

> With Magnus, the fascinating thing is that we can put ourselves in the place of the owner of a brain. One of the great problems about consciousness is that even if we could measure, through these advancing techniques, what every single neuron in the brain is doing, we might not be able to decipher what the thought is, because we do not quite know how, as that brain grew, how its connections grew, and so on. We can do it in broad outline but not in the great detail. In Magnus, because it is engineered by ourselves, we know exactly where things are,

so when you see Magnus in operation and it is thinking of a cat it actually shows the cat on the screen, because we can decode it. That for me is the fascination of being in the position of Magnus. Philosophers find that totally unacceptable, because they will say 'Oh yes, it is just dots on a screen, you know', but it may be that all our thoughts are just dots in our head.

Most of the disagreement is about whether scientists should be there at all. The philosophers and others who actually come at it from a strong religious perspective who are scientists and mathematicians would say that science is just too earthly to deal with this rather godly thing called consciousness. Now, some of the philosophers working in this area will not agree that it is godly, but will argue that it is inexplicable, therefore you cannot do science on it. But that is an after-dinner conversation which does not help the Alzheimer's patient whose consciousness has gone awry.

There is a fear of simplification at all levels. The philosopher is worried about the scientist doing any work at all and will use the word 'reductionist'. The scientist will accuse me of being outrageously reductionist in trying to build some simple systems which give you an entry into understanding much more complex systems. So there is this kind of argument all along. But I think that is not helpful. I think it would be far more helpful if people got together and said 'OK, if we reduce things a little, what is it that we actually learn, and what is it that we do not learn through this process?' rather than simply using words like 'reductionist', meaning it is no good at all.

The basic guess, which is becoming a little better focused now that I am working closer with people in neurobiology, is that consciousness is a system with loads and loads of cells in it – with lots of neurons like the brain, or artificial neurons as in Magnus – but consciousness arises as a result of some prerequisite. Some very important things have to happen in order for a particular group of neurons to contribute to a sensation of conscious experience.

Of course some engineers and scientists would say 'You can never build a machine which is big enough to represent that', or 'You can never have enough of these things in your brain, you can never have enough neurons'. They were missing a fundamental principle, and it is very simple. It goes like this:

If I have three switches or three neurons, they can fire in two to the three ways, which is eight ways. If I have four, it is sixteen. If I have a hundred, which is a very tiny brain, these things that I can represent are like two to the hundred, which is like ten and thirty noughts after it. Now if we have ten billion of these cells in our brain,

> all we need to postulate is that a very small proportion of them should be active in the business of consciousness.
>
> We have everything we want, we have a lifetime of movies in our head, which is not a technological problem. And so through the basic guess and engineering, it did seem that working with these states of firing neurons was where one would find the kind of richness that scientists often say that computers do not have.

Igor Aleksander is optimistic that forty years from now, in the same time span that it has taken computer technology to get where it is today, we will have artificially conscious machines. I am not sure I welcome that moment, but Igor Aleksander argued that this does not mean the end of humanity's long quest to discover the answer to the oldest riddle – what makes us us?

> I do not actually believe that is going to be the case, that at some point we say 'Right, here is consciousness in a box, we've done it'. I think it is more that we will develop a saner attitude towards the study of consciousness, that we will concentrate on where it matters, for example in mental illness, and concentrate on those aspects of consciousness which are distorted by mental illness. But I think one of the realisations might be that we will know what it is like to become conscious, and that in fact

was, philosophically, the beginning of the study of consciousness.

In 1690 John Locke was bored with Cartesian ideas of the pineal gland and the mind and the body having totally independent lives, or slightly dependent lives, and he said 'Well, OK, let's take a look at the mind and see how the mind becomes what it is'. And I think we are now getting back to finding the physical correlates of how that happens, but that will not put the mind in a box so we can say 'We've cracked it!' I think it will be more of an extended kind of continuous investigation where different aspects of consciousness come under scrutiny. But we are not going to be ashamed to say 'We are studying certain aspects of consciousness'. Now we are just adding to a volume of knowledge in that area, like adding to a volume of knowledge about history.

Although John Maynard Smith claimed that trying to predict the future is a mug's game, he is prepared to make some guesses.

I do know what the big unsolved problems in my own area are, that one might hope to make progress with. One, oddly enough, is the origin of life. It is a bridge between chemistry and biology. And that I am rather hopeful about because, although we could not do it in a test tube, so to

> speak, enormous progress has been made in the last fifty years, and particularly during the last twenty years. It is becoming an experimental branch of science and that is exciting. We are so much further on now than we were even ten or twenty years ago, but my impression is that this is a problem that is going to be solved in the next twenty or thirty years – the difficulties are mainly technical and chemical. I mean, they are hard to explain in non-chemical language, and let me not pretend that I understand them, because I am no chemist.

So what particular thing happened for this thing called 'life' to generate, I asked him.

> The crucial thing to happen is the origin of heredity, the origin of objects which not only replicate, so that one object gives rise to two and two gives rise to four and so on, but when they replicate they transmit their characteristics to the daughters. Now we have molecules that will do that. We can do it in a test tube, no problem. The trouble is we made the molecules. The question is, how could those molecules have arisen by sort of random collisions of other molecules? At the moment we do not fully know.

John Maynard Smith has posed the question 'Why do we have sex in a Darwinian world – it seems rather an inefficient way to carry on the purpose of the gene?'. I wanted to know whether he thought we were any nearer answering that question.

> Oddly enough, our problem now is, I would say, we have got too many answers and we are not quite sure which of them is right. And of course several of them may be right. There may be more than one reason – probably is. So our problem with the origin and maintenance of sex was rather too many different hypotheses, many of which are really very plausible. The problem is, since we were not there, it is hard to sort out which were the most important in the origin of sex. Because it was a unique event. I mean, sex, as we know it, only originated once.

Like science, according to Lewis Wolpert. Could they be related? I asked him if in, as it were, a Darwinian future, sex as we practise it would have any purpose at all. John Maynard Smith's answer was pleasantly unexpected.

> Fun. I hope we do not do away with it, it would make the world a poorer place. Look, I am sure if it were socially desired that we should, as a species, cease using the present methods of producing new kids and do it by other means, I am sure it is technically possible. I cannot imagine why we should want to, except in the very special

> cases of people who cannot have children by the normal method and are passionately desirous of having them.

If scientists can look into the not-too-distant future and believe they can solve the riddle of consciousness and the origins of life, how near are they to the elusive search for a 'theory of everything' which explains how the universe began? Paul Davies is optimistic that such a theory exists, but we are not as near as we think we are to finding the answer.

> I think there is a touch of millenarianism about this search for a Theory of Everything, or a TOE as it is sometimes called. About ten years ago there was great hope that we were finding deep linkages between different aspects of physics. That maybe in about fifty years' time we would bring it all together into a wonderful formula, a succinct mathematical description, which would encapsulate the whole of physical reality at least at some reductionist level. This would be a formula you could wear on your T-shirt at parties and say 'That's the formula for the universe'.
>
> I think, probably, we are not going to get there in fifty years. A hundred years ago there was much the same feeling, that physics was somehow drawing to a close. My own feeling about this is that such a formula probably does exist, it is probably out there

somewhere, but it is going to take us a lot longer than fifty years to find it. I would hate to think that science is a totally open-ended process – at least, this type of science.

It is very important to realise that when physicists talk about a Theory of Everything they do not mean literally everything. If we had this theory it would not explain why people fall in love, or how they vote in elections, or even about the origin of life or the nature of consciousness. It would not tell us about any of those things.

It would tell us, I think, why the world is made of the things it is; what the fundamental building blocks are; what the basic forces of nature are and how they relate to each other. It would be a theory that would unify all of the stuff, so to speak, out of which the world is put together, and I would expect space and time as well to be incorporated in that description. That would be the end of a glorious programme of work which was initiated by the Ancient Greeks, who believed that the world is nothing but atoms moving in the void, and it would be merely a matter of classifying those atoms, understanding how many there were, what their shapes were and how they link together, and you would then, in principle, have a description of everything.

Well, two and half thousand years on, I sometimes think that we are getting close to

> identifying that bottom level of reality. It is not atoms, not what we call atoms today, they are composite bodies with bits and pieces inside them, but we have broken atoms apart and we seem to have identified particles which, if they are not truly fundamental, at least I would expect are one step away from being truly fundamental. By 'truly fundamental' I mean that these would be primitive, indecomposable entities; they would not be made out of anything, they would be the bottom level of description. That would be the triumph of the reductionist programme after two and a half thousand years or so, but it would leave plenty more, plenty more for us to try to explain.

Paul Davies' conclusion that there will still be much to explain even if we discover the Theory of Everything is an optimistic one for science. But will the scientists of the future be capable of discovering the answers to the questions we can only dream about? Where is the next Newton, Einstein or Galileo? I wanted to end this series about the great scientists of the past by asking where the giants of the future are. I asked John Maddox to give me his last comment for this book from his vantage point as a close observer, assessing, editing and publishing a whole range of other scientists for the last twenty years.

> I think it was easier in the old days for people to seem giants because the universe was so badly

> known and explored. Nowadays everyone wants to be a giant and yet I would like to qualify that remark attributed to Newton by saying 'I see further because I have stood on the shoulders of giants *and pygmies*'. I think it is crucial that in modern science the contributions of quite humble people to the corporate understanding of what the world is like is just as important as the contribution of the big guys. I do believe that there are many occasions in the past fifty years when one can see how a relatively small contribution by a person who does not win a Nobel Prize, who does not get lauded around the streets of New York, is in fact of crucial importance to the development of science. It is one of the moving things about the scientific enterprise now that it depends crucially on the contributions of the pygmies as well as the giants.

There is a comparison that I can draw from my own experience as a novelist. People who write imaginative literature can, if they are not careful, be crushed by the past. Nobody is ever going to be better than Shakespeare; it is highly unlikely that anybody is going to be within touching distance of George Eliot, and so on. And to a certain extent I think that has an effect on present-day writers – and there are some very good writers indeed, but it is there, it has an effect. Now writing is a different enterprise from science because there is no 'progress' or a markedly different sort of development, but I wondered if John Maddox thought that it was slight crushing for young scientists now to be faced by these enormous achievers, gigantic leaps forward, quite colossal contributions to science.

> Well, within our own lifetimes DNA was discovered. In every sort of way that discovery, which would have been made by somebody within three or four years of the point at which Crick and Watson made it, ranks with anything else in the past in its importance, in its value as a stimulus to other scientific research, and as a way of casting the shades from our own eyes in our contemplation of what the world is like. I think it is certainly a discovery as important as Copernicus' putting the sun and not the earth at the centre of the world.
>
> So in that spirit there are still discoveries to be made. It seems to me that the people who are working hard on trying to understand the brain will, in due course, make discoveries that surprise us all and which turn out to be exceedingly important. It seems to me that the same is true with the origin of life, the point at which somebody is able to take some chemicals, put them in a pot, heat them up, perhaps, and then, a few days later, see that little organisms are living in his pot. That man will be a very happy, delighted fellow – and it may be a woman, of course. It seems to me, therefore, that at the coal face, in the laboratory, there is still plenty of opportunity for making one's

> name. No, I do not think the gilt has gone off the gingerbread at all. I think it is still a marvellous enterprise.

Scientists seek truth through knowledge and progress comes through success in this 'marvellous enterprise'. There is also a more extreme, perhaps more romantic view.

We are told that the universe came into existence about fifteen billion years ago with the Big Bang. On our earth, for most of the four thousand million years it has been in existence, there was no living creature or thing. If we equate the age of the earth to a twenty-four-hour day, the first signs of life appear after the twenty-third hour and human beings emerge in the last few minutes before midnight.

This analogy of the clock is often used. It seems to me to carry a fatal pessimism. For when midnight strikes – is that not the Apocalypse, the end of everything? Why could our few minutes not be the first of another fifteen billion year adventure?

Because – as we have seen – in a mere hundred generations, since the Greeks, the scientific component of the human brain has unleashed itself from superstition and ignorance and is now launched on an astonishing mission whose purpose, it seems to me, is no less than to seek out its Maker.

FURTHER READING

Archimedes

Geoffrey Lloyd, 'New Perspectives on Ancient Science', *European Review*, 1994

Geoffrey Lloyd, *Early Greek Science: Thales to Aristotle*, Chatto & Windus, 1970

Lewis Wolpert, *The Unnatural Nature of Science*, Faber & Faber, 1992

John Fauvel and Jeremy Gray (eds), *The History of Mathematics: A Reader*, Macmillan, in association with The Open University, 1987

Lisa Jardine, *Worldly Goods: A New History of the Renaissance*, Macmillan, 1996

Galileo Galilei

Mario Biagioli, *Galileo Courtier: The Practice of Science in the Culture of Absolutism*, University of Chicago Press, 1993

Michael Sharratt , *Galileo, Decisive Innovator*, Cambridge University Press, 1996

Margaret Wertheim, *Pythagoras' Trousers: God, Physics and the Gender Wars*, Fourth Estate, 1997

Paul Davies, *The Mind of God: Science and the Search for Ultimate Meaning*, Simon and Schuster, 1992

Sir Isaac Newton

Robert Iliffe, '"Is He Like Other Men?": The Meaning of the Principia Mathematica', and 'The Author as Idol', from *Culture and Society in the Stuart Restoration*, ed. Gerald Maclean, Cambridge University Press 1995

Richard S. Westfall, *Never At Rest: A Biography of Isaac Newton*, Cambridge University Press, 1980

A. Rupert Hall, *Isaac Newton: Adventure in Thought*, Cambridge University Press, 1996

Michael White, *Isaac Newton: The Last Sorcerer*, Fourth Estate, 1997

Antoine Lavoisier

Arthur Donovan, *Antoine Lavoisier: Science, Administration and Revolution*, Cambridge University Press, 1996

Jean-Pierre Poirier, *Lavoisier: Chemist, Biologist, Economist*, University of Pennsylvania Press, 1996

Bernadette Bensaude-Vincent and Isabelle Stengers, *A History of Chemistry*, Harvard University Press, 1996

Simon Schaffer, *Science, Medicine and Dissent*, eds. F.R.G. Anderson and Christopher Lawrence, Wellcome Institute, 1987

Michael Faraday

John Meurig Thomas, *Michael Faraday and the Royal Institution: The Genius of Man and Place*, Adam Hilger, 1991

Geoffrey Cantor, 'Michael Faraday: Sandemanian and Scientist' in *A Study of Science and Religion: I – The Nineteenth Century*, Macmillan, 1991

Michael Faraday, *Selected Correspondence*, vols. I and II, Cambridge University Press, 1971

Charles Darwin

Janet Browne, *Charles Darwin* vol. I: *Voyaging*, Jonathan Cape, 1995

Adrian Desmond and James Moore, *Darwin*, Michael Joseph, 1991

Stephen Jay Gould, *Ever Since Darwin: Reflections in Natural History*, Penguin Books, 1978

Stephen Jay Gould, *Hens' Teeth and Horses' Toes: Further Reflections in Natural History*, Penguin Books, 1984

Daniel Dennett, *Darwin's Dangerous Idea: Evolution and the Meanings of Life*, Allen Lane, The Penguin Press, 1995

John Maynard Smith, *The Theory of Evolution*, Cambridge University Press, 1993

Charles Darwin, *The Beagle Diary*, edited by **R. D. Keynes**, Cambridge University Press, 1988

Jules Henri Poincaré

June Barrow Green, 'Poincaré and the Three-Body Problem', in *History of Mathematics*, vol. II, American Mathematical Society, 1997

Ian Stewart, *Nature's Numbers: Discovering Order and Pattern in the Universe*, Weidenfeld and Nicolson, 1995

Henri Poincaré, *Science and Method*, Thoemmes Press, 1996

Sigmund Freud

Adam Phillips, *On Kissing, Tickling and Being Bored*, Faber & Faber, 1993

Susan Greenfield *Journeys to the Centres of the Mind: Towards a Science of Consciousness*, W. H. Freeman, 1995

Oliver Sacks, 'Neurology and the Soul', in *The Enchanted Loom: Chapters in the History of Neuroscience*, ed. Pietro Corsi, Oxford University Press, 1991

Peter Gay, *Freud: A Life for our Time*, Dent and Sons, 1988

Sigmund Freud, *Case Histories I: Dora and Little Hans*, Pelican Books, 1977

Ernest Jones, *The Life and Work of Sigmund Freud*, Hogarth Press, 1953–7

Marie Curie

Eve Curie, *Madame Curie*, Doubleday, 1937

Susan Quinn, *Marie Curie: A Life*, Heinemann, 1995

Albert Einstein

Paul Davies, *About Time: Einstein's Unfinished Revolution*, Viking, 1995

Abraham Pais, *Einstein Lived Here*, Oxford University Press, 1994

Abraham Pais, *Subtle Is The Lord*, Clarendon Press, 1982

Ronald Fraser, *Einstein: The Life and Times*, Hodder & Stoughton, 1973

Stephen Hawking and **Roger Penrose**, *The Nature of Space and Time*, Princeton University Press, 1996

Martin J. Klein. A.J. Kox, Jurgen Renn and **Robert Schulmann**, *The Collected Papers of Albert Einstein*, Princeton University Press, six volumes, to 1996

Francis Crick and James Watson

James Watson, *The Double Helix: A Personal Account of the Discovery of the Structure of DNA*, Weidenfeld and Nicolson, 1997 (first published 1968)

Francis Crick, *What Mad Pursuit: A Personal View of Scientific Discovery*, Penguin Books, 1990

Evelyn Fox Keller, *Refiguring Life: Metaphors of Twentieth-Century Biology*, Columbia University Press, 1995

Peter Medawar, *The Strange Case of Spotted Mice*, Oxford University Press, 1996

Where Are We Now?

John Horgan, *The End of Science: Facing the Limits of Knowledge in the Twilight of the Scientific Age*, Little, Brown and Co., 1996

John Maddox, *What Remains to be Discovered*, Macmillan, 1998

Igor Aleksander, *Impossible Minds: My Neurons, My Consciousness*, Imperial College Press, 1996

Roger Penrose, *The Emperor's New Mind*, Vintage, 1990

Richard Dawkins, *The Selfish Gene*, Oxford University Press, 1989

ACKNOWLEDGMENTS

Extract from *Lavoiser: Chemist, Biologist, Economist* by Jean-Pierre Poirier. Translated by Rebecca Balinski. Copyright © 1996 University of Pennsylvania Press. Reprinted by permission of the publisher.

Extract from *Arcadia* by Tom Stoppard. Copyright © 1993 by Tom Stoppard. Reprinted by permission of Faber and Faber, Inc.

Extracts from *Marie Curie: A Life* reprinted with the permission of Simon & Schuster, from *Marie Curie* by Susan Quinn. Copyright © 1995 by Susan Quinn.

Extracts from *Letters of Sigmund Freud* 1873–1939, edited by Ernst L. Freud. Copyright © 1960 by Sigmund Freud Copyrights, Ltd. Copyright renewed. Reprinted by permission of Basic Books, a member of Perseus Books, L.L.C. For additional rights/territory contact Sigmund Freud Copyrights, Ltd., 10 Brook Street, Wivenhor, Colchester CO7 9DS England.

Extract from *The Double Helix* reprinted with the permission of Scribner, a division of Simon & Schuster, Inc., from *The Double Helix* by James D. Watson. Copyright © 1968 James D. Watson.

Extracts from *On the History of the Psycho-Analytic Movement* by Sigmund Freud, translated by Joan Riviere. Translation copyright © 1962 by James Strachey. Reprinted by permission of W. W. Norton & Company, Inc.

Extract from *The Race for the Double Helix* © William Nicholson, published by Horizon.

Extract from *Makers of Rome: Nine Lives* by Plutarch, translated by Ian Scott-Kilvert (Penguin Classics 1965) © Ian Scott-Kilvert 1965. Reproduced by permission of Penguin Books.

Every effort has been made to trace copyright holders in all copyright material. The publisher regrets if there has been any oversight and suggests that John Wiley & Sons be contacted in any such event.

Illustrations

Mary Evans Picture Library: pp. 11, 101, 157, 213, 243, 273, 301. Hulton-Getty, p. 43. Science Library/Science & Society Pic Lib: pp. 73, 127. Roger Viollet, Paris: p. 185.